Fracking

Scrivener Publishing
100 Cummings Center, Suite 541J
Beverly, MA 01915-6106

Publishers at Scrivener
Martin Scrivener (martin@scrivenerpublishing.com)
Phillip Carmical (pcarmical@scrivenerpublishing.com)

Fracking

The Operations and Environmental Consequences of Hydraulic Fracturing

Michael D. Holloway

& Oliver Rudd

Scrivener Publishing

WILEY

Co-published by John Wiley & Sons, Inc. Hoboken, New Jersey, and Scrivener Publishing LLC, Salem, Massachusetts.
Published simultaneously in Canada.

For general information on our other products and services or for technical support, please contact our Customer Care Department within the United States at (800) 762-2974, outside the United States at (317) 572-3993 or fax (317) 572-4002.

Wiley also publishes its books in a variety of electronic formats. Some content that appears in print may not be available in electronic formats. For more information about Wiley products, visit our web site at www.wiley.com.

For more information about Scrivener products please visit www.scrivenerpublishing.com.

Cover design by Kris Hackerott

Library of Congress Cataloging-in-Publication Data:

ISBN 978-1-118-49632-9

Printed in the United States of America

10 9 8 7 6 5 4 3 2

Dedication

…for the children

Contents

Preface

Hydraulic fracturing (also known as fracking, fracing, or worse) of rock deep beneath the surface to release petroleum product has become a contentious subject across the globe. This practice is not to be confused with drilling or extraction. Fracking is the process of using fluid power to fracture rock to release gas (and sometimes crude oil). It is not drilling per se, although drilling must be done to establish a well in order to pump fluid that fractures rock to release product. Certain countries have actually outlawed the practice of hydraulic fracturing, claiming that ground water and air pollution increase due to the practice and that the comfort of life is adversely affected. Legitimate concerns are always available, and examples to purport a concerned view are magnified. The media plays a large role in acting as a magnifier instead of an information conduit. While a magnifying glass produces a very close examination, it suffers blurred vision when used to examine the landscape. The intent of this work is to provide a correct and balanced view of fracturing underground rock with fluids in order to release a product to produce energy. The concept of using water to do work is nothing new. Pumping fluid below ground in order to fracture rock in order to release gaseous petroleum is a new process. The practice is done with surprising precision as well as environmental concern, yet it is interesting how the public reacts to the practice in relation to other techniques used throughout the world. This work will explore the materials used as well as the concerns most common to the practice.

The primary use of this book shall be to increase the awareness of new and emerging technologies and what the various ramifications can be. The reader shall be exposed to many engineering concepts and terms. All of these ideas and practices shall be explained within the body. A science or engineering background is not required. Technical and non-technical readers will be able to use this book for many reasons: research, investment information, a litigation resource, political research for public policy, and as an educational instrument. The series contents include the history, public and media perception, products produced, methodology, materials used, reserves and production, environmental impact, and economic impact. Energy companies are encouraged to use this work as a means to educate the general population. Please feel free to use any, if not all, of the content that follows, provided copyright laws are followed.

Introduction

When people say that one cannot squeeze blood out of a turnip, it means that one cannot get something from a person, especially money, that the person does not have. One cannot squeeze blood from a turnip because it is basically impossible. Many years ago, who would have thought that methane could be obtained from rock? Doing so would be right in line with squeezing blood from a turnip. In actuality, trapped natural gas can be released from rock; at least, that is what is being accomplished now throughout North America and others parts of the world. One hundred years ago, no one thought it possible. Very few, if any, contemplated the idea. Natural gas, which is primarily methane, has been proven to be an excellent fuel source. It can be safely burned to create heat to power engines, boilers in factories and homes, as well as powering turbines for generating electricity. Projections on natural gas volumes trapped underground suggest a near inexhaustible supply of this product, yet such abundance spawns controversy. A popular and economical technique that relies on the gas from subterranean sources requires fracturing rock bed. This process is actually carried out naturally every day with water or magma. Magma may flow into rock beds, superheating water to generate steam. The resulting pressure of the expanding water molecule can be so great it can lift and separate thousands of tons of rock deep beneath the earth's surface. This same practice can be carried out artificially (induced) using high-powered pumps and various liquid compounds. This technique, combined with new horizontal directional drilling machines, has enabled the harvest and distribution of natural gas. But at what cost? Does this practice contribute to greenhouse gas? Does it create earthquakes? Does it contaminate the groundwater supply? These are important ideas to consider yet, and with proper examination and logic, the authors are confident that the reader will gain insight and reason in a practice fueled by profit and civil concern.

The following list contains excellent work used either to support fracking or used as a tool to oppose it.

Studies Used to Support Hydraulic Fracturing

Interstate Oil and Gas Commission "States' Experience with Hydraulic Fracturing" 2003

U.S. EPA "Evaluation of Impacts to Underground Sources of Drinking Water by Hydraulic Fracturing of Coalbed Methane Reservoirs" 6/2004

U.S. Department of Energy and Ground Water Protection Council "Modern Shale Gas Development in the United States: A Primer" 4/2009

U.S. Department of Energy and Ground Water Protection Council "State Oil and Gas Regulations Designed to Protect Water Resources" 5/2009

IF International "Analysis of Subsurface Mobility of Fracturing Fluids" 2009

NY State Department of Environmental Conservation "DRAFT Supplemental Generic Environmental Impact Statement" 9/2009

MIT Energy Initiative "The Future of Natural Gas: An Interdisciplinary MIT Study, Interim Report" 9/2010

Studies Used to Oppose Hydraulic Fracturing

Oil & Gas Accountability Project of EARTHWORKS "Our Drinking Water at Risk: What EPA and the Oil and Gas Industry Don't Want Us to Know about Hydraulic Fracturing" 4/2005

Natural Resources Defense Council "Drilling Down: Protecting Western Communities from the Health and Environmental Effects of Oil and Gas Production" 10/2007

Southern Methodist University for Environmental Defense Fund "Emissions from Natural Gas Production in the Barnett Shale Area and Opportunities for Cost–Effective Improvements" 1/2009

NYC Department of Environmental Protection and Hazen and Sawyer "Impact Assessment of Natural Gas Production in the New York City Water Supply Watershed" 12/2009

Harvey Consulting LLC Review of NY State's Draft Supplemental Generic Impact Statement 12/2009

Texas Commission on Environmental Quality "Health Effects Review of Barnett Shale Formation Area Monitoring Projects" 1/2010

Valid points are made on both fronts. The major concern against fracking resides in the overall health and well-being of people close to a well site, as well as the land, water, and air that could potentially be adversely affected. It is highly advisable that the reader obtain his own copies of these reports if he wishes to delve deep into the dark recesses of this topic.

The issue of hydraulic fracturing is one of the most contentious and misunderstood processes, even down to its name (is it frac or frack?). The aim of this book is not to deliver a highly technical level explanation of the process or to lay out results of study after study that have been conducted for both sides of the subject…those books have been written time and time again and, quite honestly, someone reading those books has for the most part made up his mind on fracking and knows how he feels on the subject.

No sides are taken in the following pages. This work is not intended to be pro-industry or anti-fracturing. Namely, it aims to educate the general public on what hydraulic fracturing really is, how it is conducted, and what possible harms may or may not come as a result. Think of this as "Hydraulic Fracturing 50 ½" in preparation of taking "Hydraulic Fracturing 101" and becoming a better informed individual on the subject. Both sides are equally welcomed to enjoy this book…the only hope is that both are more educated, and hopefully at least a little bit entertained, for reading it.

1

Environmental Impact – Reality and Myth and Nero Did Not Fiddle While Rome Burned

In today's society, it is really easy for organizations – be it the general media, political groups, local organizations, unions, or religious associations – to spread their beliefs to the public and push whatever agenda or ideals they may have. These beliefs could be successfully put forward with good intentions, successfully put forth with bad intentions, or, in many cases, put forward with good intentions, but have a negative result. Sadly, it seems human nature dictates that the first opinion heard or the opinion heard the loudest and with the most hyperbole will be what the public comes to believe. In time, once something is believed by enough people and stated as "fact" long enough, the general public will no longer even bother looking into facts, and it will become part of the fabric of beliefs in our society – for instance, a few examples of this phenomenon are: 1) one, in fact, cannot see the Great Wall of China from the moon (not even close); 2) the Sherlock Holmes character never once said "Elementary, my dear Watson"; and 3) Nero didn't play the violin while his city burned. The violin wasn't invented yet.

Now, upon reading this, the hope is that many will stop reading (but of course come back) to fact-check that what is written here is true. That is exactly the point to push: as previously stated many times, the point is not to give opinions or try to sway beliefs, the point is to merely give facts described in the most straightforward and clear way possible and hopefully give enough background knowledge to initiate further study. Following this and further study, hopefully one can make up his or her own educated opinion on fracking…and that Nero actually funded relief money from his own savings to help his subjects.

As far as hydraulic fracturing is concerned, the aspect given the most attention by press and most concerned organizations is the impact it may have on the environment. The question of environmental impact through fracking is, to say the least, a very emotional topic and by far the most polarizing issue; however, a great deal of analysis indicates that the most significant environmental risks attributed to fracking are similar to risks long associated with all drilling operations, including groundwater contamination due to inadequate cementing and/or well construction, risks associated with trucking, leaks from tanks and piping, and spills from waste handling. This all-encompassing blame has given industry all of the ammunition needed to claim that effects attributed to hydraulic fracturing are overstated, not based on good science, or related to processes other than hydraulic fracturing.

Due to the great ongoing controversy over alleged impacts from fracking, many public groups have become deeply suspicious of the trustworthiness and overall motives of the oil and gas industry. These suspicions are continuously intensified by two things:

1. ongoing mistrust of data and findings due, in great part, to semantics, and
2. by the industry initially refusing to disclose the chemical makeup of fracking fluids and the additives used to enhance hydraulic fracturing.

1.1 The Tower of Babel and How it Could be the Cause of Much of the Fracking Debate

Almost everyone has heard the story, or has a general understanding, of the Tower of Babel from the Old Testament. In the Biblical account of this story, humanity was attempting, as a unified group, to build a tower in Mesopotamia to reach the heavens, only to have their efforts brought to a halt by one of the most effective means imaginable. The efforts of this united group of people were not thwarted by military force, or by weather, or even by sickness and injury; their efforts in this undertaking were thwarted by speech. The simple fact of this story is that building of the tower came to a halt once the unified people were confounded by speech and no longer able to communicate to work together. Now far be it for this work to compare modern day hydraulic fracturing with the construction, and subsequent stop in construction, of the Tower of Babel, but much of the confusion, name-calling, and general mistrust between groups on this subject can be attributed to a difference in communication. Maybe once this communication gap is bridged, more effective talks can be established between industry and the concerned public in place of wasting time on mistrust and name-calling. Hopefully, this work can help to bridge that gap.

It can be easily considered that a very large portion of negativity toward hydraulic fracturing is actually attributable to processes other than hydraulic fracturing. In the discussions between industry and the public, a great deal of this problem can boil down to an issue of semantics: the oil and gas industry has a narrow view of what fracking entails (including just those processes related to the actual process of fracking while on location conducting the fracking operation), while the general public is more inclined to include many more activities commonly related to fracking (water and sand trucking, product and equipment transport and storage, water disposal) under the heading of "fracking." This can cause misunderstandings and skewed data, in that many of the processes included by the general public are utilized in many, if not all, drilling practices, and are hard to put solely under the heading of "fracking," when in actuality they could just as easily be under the heading "completions" or "production."

This topic has been discussed many times in the media, in town hall meetings, as well as on various concerned citizens' blogs to the point of exhaustion. There are many proven environmental impacts caused by drilling operations and processes related to drilling. There are also many concerns raised about any industrial effort. Any time that man and machine are working, there are countless opportunities for an environmental issue to emerge. It is the nature of the beast. It has been suggested by many that the advent of industry in England as well as the northeast of the United States changed the seasons, the combination of drought and unmanaged farming led to the Midwestern Dust Bowl of last century, and the increase in exhaust emissions from automobile exhaust has led to unsafe levels of pollutants in all cities. These are facts. Facts can almost assuredly not be disproven by industry, and can be a concern by the public in their feelings on gas well completion and production activities. However, by the same token, there have been over a million wells that have gone through the process of hydraulic fracturing. With such a great opportunity for environmental impact, one would think that the process would surely have poisoned the water. With the current issue of semantics, public concerns can include many drilling processes, while industry can fall back on the fact that the industry definition of fracking has never impacted fresh water in the ways commonly claimed by the media for public consumption. The debate can rage on, with both sides being right and both sides being wrong, while never taking steps to come together on a common goal. The goal of this work is not to shine light on the mistakes of the drilling industry or show how citizens can make false claims; the goal is to educate and share insight. If the one message that can be taken away from this work is how to drive this technology safely, then the goal shall be reached.

2

Production Development

Before an investigation is begun into the production of a well site, it is important to understand what the product is. There are several types of 'product' that are obtained at the well site. Gas reservoirs are classified as conventional or unconventional for the following reasons:

Conventional reservoirs: Wells in conventional gas reservoirs produce from sand carbonates (lime stones and dolomites) that contain the gas in interconnected pore spaces that allow flow to the wellbore. Gas in the pore scan will move from one pore to another through smaller pore-throats that create permeable flow through the reservoir. In conventional natural gas reservoirs, the gas is often sourced from organic-rich shale proximal to the more porous and permeable sandstone or carbonate.

Unconventional reservoirs: Wells in unconventional reservoirs produce from low permeability (tight) formations such as tight sands and carbonates, coal, and shale. In unconventional gas reservoirs, the gas is often sourced from the reservoir rock itself (tight gas sandstone and carbonates are an exception). Because of the low permeability of these formations, it is typically necessary to stimulate the reservoir to create additional permeability. Hydraulic fracturing of a reservoir is the preferred stimulation method for gas shale. Differences between the three basic types of unconventional reservoirs include:

- **Tight Gas:** Wells produce from regional low-porosity sandstones and carbonate reservoirs. The natural gas is sourced (formed) outside the reservoir and migrates into the reservoir over time (millions of years). Many of these wells are drilled horizontally, and most are hydraulically fractured to enhance production.
- **Coal Bed Natural Gas** (CBNG): Wells produce from the coal seams, which act as source and reservoir of the natural gas. Wells frequently produce water as well as natural gas. Natural gas can be sourced by thermogenic alterations of coal or by biogenic action of indigenous microbes on the coal. There are some horizontally drilled CBNG wells, and some that receive hydraulic fracturing treatments. However, some CBNG reservoirs are also underground sources of drinking water, and as such, there are restrictions on hydraulic fracturing. CBNG wells are mostly shallow, as the coal matrix does not have the strength to maintain porosity under the pressure of significant overburden thickness.
- **Shale Gas:** Wells produce from low permeability shale formations that are also the source for the natural gas. The natural gas volumes can be stored in a local macro-porosity system (fracture porosity) within the shale, or within the micro-pores of the shale or it can be adsorbed onto minerals or organic matter within the shale. Wells may be drilled either vertically or horizontally, and most are hydraulically fractured to stimulate production. Shale gas wells can be similar to other conventional and unconventional wells in terms of depth, production rate, and drilling.

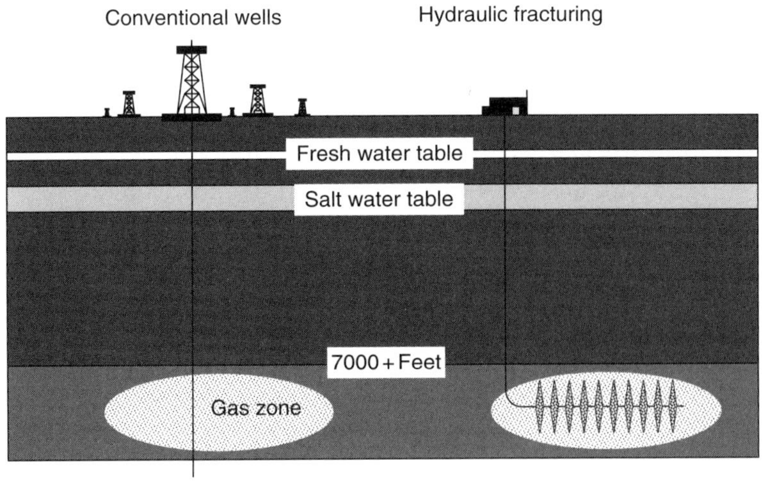

Figure 2.1 Comparison of Well Sites.

The following is a comparison of conventional well structure and hydraulic fracturing.

Drilling conventional wells and those to be used in hydraulic fracturing begin in much the same ways. The basic well construction steps are:

- An initial length of steel pipe, called conductor casing, is inserted into a vertical wellbore soon after drilling begins. This is done to stabilize the well as it passes through the shallow sediments and soils near the surface.
- Once conductor casing is set, operators continue drilling and insert a second casing, called surface casing, from the ground surface and extending past the depth of all drinking water aquifers.
- After allowing the cement behind the casings to set (cementing is described in detail in the following section), operators continue drilling for approximately 10 to 50 feet before stopping to test the integrity of the cement process by pressurizing the well.
- In horizontal wells, after drilling the horizontal section of the well, operators run a string of production casing into the well and cement it in place.
- Operators then perforate the production casing using small explosive charges at intervals along the

horizontal wellbore where they intend to hydraulically fracture the shale.

- Acid stage; consisting of several thousand gallons of water mixed with a dilute acid such as hydrochloric or muriatic acid: This serves to clear cement debris in the wellbore and provide an open conduit for other frack fluids by dissolving carbonate minerals and opening fractures near the wellbore.
- Pad stage; consisting of approximately 100,000 gallons of slickwater without proppant material: The slickwater pad stage fills the wellbore with the slickwater solution (described below), opens the formation, and helps to facilitate the flow and placement of proppant material.
- Prop sequence stage; which may consist of several sub-stages of water combined with proppant material (consisting of a fine mesh sand or ceramic material, intended to keep open, or "prop," the fractures created and/or enhanced during the fracturing operation after the pressure is reduced): This stage may collectively use several hundred thousand gallons of water. Proppant material may vary from a finer particle size to a coarser particle size throughout this sequence.
- Flushing stage; consisting of a volume of fresh water sufficient to flush the excess proppant from the wellbore.

Other additives commonly used in the fracturing solution employed include:

- A dilute acid solution, as described in the first stage, used during the initial fracturing sequence. This cleans out cement and debris around the perforations to facilitate the subsequent slickwater solutions employed in fracturing the formation.
- A biocide or disinfectant, used to prevent the growth of bacteria in the well that may interfere with the fracturing operation: Biocides typically consist of bromine-based solutions or glutaraldehyde.

- A scale inhibitor, such as ethylene glycol, is used to control the precipitation of certain carbonate and sulfate minerals.
- Iron control/stabilizing agents such as citric acid or hydrochloric acid, are used to inhibit precipitation of iron compounds by keeping them in a soluble form.
- Friction reducing agents, also described above, such as potassium chloride or polyacrylamide-based compounds, used to reduce tubular friction and subsequently reduce the pressure needed to pump fluid into the wellbore: The additives may reduce tubular friction by 50 to 60%. These friction-reducing compounds represent the "slickwater" component of the fracking solution.
- Corrosion inhibitors, such as N,n-dimethyl formamide, and oxygen scavengers, such as ammonium bisulfite, are used to prevent degradation of the steel well casing.
- Gelling agents, such as guar gum, may be used in small amounts to thicken the water-based solution to help transport the proppant material.

Occasionally, a cross-linking agent will be used to enhance the characteristics and ability of the gelling agent to transport the proppant material. These compounds may contain boric acid or ethylene glycol. When cross-linking additives are added, a breaker solution is commonly added later in the frack stage to cause the enhanced gelling agent to break down into a simpler fluid so it can be readily removed from the wellbore without carrying back the sand/proppant material.

3

Fractures: Their Orientation and Length

Certain predictable characteristics or physical properties regarding the path of least resistance have been recognized since hydraulic fracturing was first conducted in the oilfield in 1947. The following table lists various sites also known as Plays throughout North America.

3.1 Fracture Orientation

Hydraulic fractures are formed in the direction perpendicular to the least stress. Based on experience, horizontal fractures will occur at depths less than approximately 2000 ft because the earth's overburden at these depths provides the least principal stress. If pressure is applied to the center of a formation under these relatively shallow conditions, the fracture is most likely to occur in the horizontal plane, because it will be easier to part the rock in this direction than in any other. In general, therefore, these fractures are parallel to the bedding plane of the formation.

Table 3.1 Well Comparisons.

Gas Shale Basin	**Barnett**	**Fayetteville**	**Haynesville**	**Marcellus**	**Woodford**	**Antrim**	**New Albany**
Estimated Basin Area, square miles	5,000	9,000	9,000	95,000	11,000	12,000	43,500
Depth, ft	8,500	7,000	13,500	8,500	11,000	2,200	2,000
Thickness, ft	To 600	To 200	To 300	To 200	To 220	To 120	To 100
Depth to Base of Treatable Water#, ft	~1200	~500	~400	~850	~400	~300	~400
Rock Column Thickness between Top of Play and Bottom of Treatable Water, ft	7,300	6,500	13,100	7650	10,600	1,900	1,600
Total Organic Carbon, %	4.5	9.8	4.0	12	14	20	25
Total Porosity, %	4 –	8	9	10	9	9	14
Gas Content, SCF/Ton	To 350	To 220	To 330	To 100	To 300	To 100	To 80
Water Production, Barrels water/day	N/A	N/A	N/A	N/A	N/A	5 – 500	5 – 500
Well spacing, acres	60 to 160	80 to 160	40 to 560	40 to 160	640	40 to 160	80
Original Gas-In- Place, TCF	327	52	717	1,500	23	76	160
Technically Recoverable Resources	44	41.6	251	262	11.4	20	19.2

As depth increases beyond approximately 2000 ft, overburden stress increases by approximately 1 psi/ft, making the overburden stress the dominant stress. This means the horizontal confining stress is now the least principal stress. Since hydraulically induced fractures are formed in the direction perpendicular to the least stress, the resulting fracture at depths greater than approximately 2000 ft will be oriented in the vertical direction. In the case where a fracture might cross over a boundary where the principal stress direction changes, the fracture would attempt to reorient itself perpendicular to the direction of least stress. Therefore, if a fracture propagated from deeper to shallower formations, it would reorient itself from a vertical to a horizontal pathway and spread sideways along the bedding planes of the rock strata.

3.2 Fracture Length/ Height

The extent that a created fracture will propagate is controlled by the upper confining zone or formation, and the volume, rate, and pressure of the fluid that is pumped. The confining zone will limit the vertical growth of a fracture because it either possesses sufficient strength or elasticity to contain the pressure of the injected fluids, or an insufficient volume of fluid has been pumped. This is important because the greater the distance between the fractured formation and the USDW, the more likely it will be that multiple formations possessing the qualities necessary to impede the fracture will occur. However, while it should be noted that the length of a fracture can also be influenced by natural fractures or faults, as shown in a study that included micro-seismic analysis of fracture jobs, natural attenuation of the fracture will occur over relatively short distances due to the limited volume of fluid being pumped and dispersion of the pumping pressure, regardless of intersecting migratory pathways.

4

Casing and Cementing

Typically, the casing and cementing of wells is accomplished in multiple phases, working inward from the from the largest diameter casing to the smallest: picture a pirate's spyglass opened up and buried with the eyepiece at the deepest point underground.

- First, as described above, the surface casing is inserted.
- Cement is then pumped down the inside of the casing, forcing it up from the bottom of the surface casing into the annular space between the outside of the surface casing and the drilled wellbore.
- Once a sufficient volume of cement to fill the annulus is pumped into the casing, it is usually followed by pumping water into the casing until the cement begins to return to the surface in the annular space. This process of cementing is called "circulation" and ensures that the entire annular space fills with cement from below the deepest ground water zone

  to the surface, which, when done properly, protects the aquifer.
- After the surface casing is completed and set, the well is drilled to the pay zone. Upon reaching the pay zone, production casing is set at either the top extent of, or into, the producing formation. This zone is then cemented by the same process as before.

As can be expected when working in underground formations, sometimes a well will be extended into a formation that is difficult to cement because of high formation porosity or high water flow. This condition can be detected by personnel during drilling operations or by observing that more cement is being pumped downhole than the calculated area of the annulus. When this happens, additives (such as cellophane flake and calcium chloride) can sometimes be added to the cement to seal off such zones, quicken the cement hardening process, and prevent downhole loss of the cement into formation.

When completed properly, sealing of annular spaces with cement creates a hydraulic barrier to both vertical and horizontal fluid migration, essentially eliminating subsurface releases and blowouts. Consequently, the quality of the initial cement job is one of the most critical factors in the prevention of fluid movement into groundwater resources. If well cementing and casing is improperly or poorly done, leakage into an aquifer of fluids that flow upward in the annulus between the casing and the borehole can occur: this is the greatest potential pollution risk.

Well integrity issues resulting in leakage can be divided into two categories:

- Annular flow: fluids move up the wellbore by traveling up the space between the borehole formation and cement or between the cement and the casing
- Leak flow: fluids move in a radial direction out of the well and into the adjacent formation.

4.1 Blowouts

The simple definition of a blowout is an uncontrolled fluid release occurring during the drilling, completion, or production of oil and gas wells. They rarely occur, however: when blowouts occur, it is typically when unexpectedly high pressures are encountered in the subsurface or due to valve or some other type of mechanical failure. What is normally brought to mind when one hears the term "blowout" is the highly graphic explosion that goes with a catastrophic well blowout; however, as used here, this term also includes the smaller above and below ground blowouts.

Blowouts may take place at the surface (wellhead or elsewhere) or subsurface (naturally high pressure, or may be artificially induced in the well bore during hydraulic fracturing during completion operations, but not during pumping). A high percentage of blowouts occur due to casing or cement failure, allowing high-pressure fluids to escape up the well bore and flow into subsurface formations. Blowouts have shown to be the most common of all well control problems, and they also appear to be underreported. The potential environmental consequences of a blowout depend mostly on:

- The timing of the blowout relative to well activities, which determines the nature of the released fluid such as natural gas or pressurized fracturing fluid;
- Occurrence of the escape of containments through the surface casing or deep in a well; and
- The risk receptors, such as freshwater aquifers or water wells that are impacted.

4.2 Surface Blowouts

Surface blowouts at the wellhead are serious matters that can result in a major safety hazard to workers and may also result in surface spills. Surface blowouts are primarily prevented through proper well construction, maintenance, and ensuring well integrity.

4.3 Subsurface Blowouts

Subsurface blowouts, due to high gas pressure or mechanical failures, happen in both regular and hydraulically fractured wellbores. However, fractured wells have the incremental risk of potential failures caused by the high pressures of fracturing fluid during the process.

In the event of a blowout in the subsurface, a major problem due to the limited ability to discern what is happening in the subsurface, blowout preventers are used to automatically shut down fluid flow in the well bores. Subsurface blowouts may pose both safety hazards and environmental risks. For example, when a blowout preventer engages to prevent flow from reaching the surface, the fluid may be forced through weaknesses in the casing and cement below the blowout preventer into the surrounding formations and aquifers. Blowout preventers are important safety devices; however, like all mechanical devices, they have been known to fail, sometimes with highly visible and catastrophic results.

4.4 Horizontal Drilling

When drilling a horizontal well, operators begin turning the drill bit when they near the production zone so the wellbore runs through the formation horizontally: then it can extend up to 10,000 feet, which vastly increases contact with the production zone relative to vertical drilling.

4.5 Fracturing and the Groundwater Debate

While hydraulic fracturing in particular has been the primary focus of controversy and numerous demonstrations, studies indicate that environmental risks associated with hydraulically fractured wells are similar to those associated with all production wells, including surface and subsurface spills and releases, gas migration and groundwater contamination due to faulty well construction, blowouts, and leaks and spills of waste water and chemicals stored on pad sites.

Of all the issues related to hydraulic fracturing, the possible effects on groundwater are without a doubt the most contentious. Numerous allegations have been made related to hydraulic fracturing, with particular emphasis on impacts to water wells. More of this serious issue will be covered in detail.

5

Pre-Drill Assessments

If fresh water aquifer contamination is a concern, many concerned parties – both in industry and in the environmental sector – believe shallow water monitoring wells should be drilled at the perimeters of the pads, with samples taken as needed. Collecting samples from these water wells before drilling would then be done to establish a base line set of analytical data to show water conditions prior to drilling operations. In fact, some states have enacted regulations to establish pre-drilling groundwater quality through a baseline monitoring program. The baseline sampling would then be followed by regular monitoring during all phases operations, including hydraulic fracturing. This process could serve two purposes:

- Provide very early warning in case of a well construction leak; and
- Eliminate the operation as a potential source of pollution in some areas of concern, as when it is shown

> that constituents were present in water wells before shale gas development began.

High-quality analytical data before, during, and after production processes have been increasingly collected to document baseline and post-drilling groundwater quality through wells installed at the well pad. However, one obvious shortcoming of this protocol is that, as stated previously in this book, typical horizontal lengths may range from 2,000 feet to 6,000 feet, with extremes of 12,000 feet or more. Therefore, if water monitoring wells are located at the well pad, they could be anywhere from 2,000 to 12,000 feet from the location of the actual fracturing process.

The following is an excellent paper written by Lloyd Hetrick from an EPA workshop and reprinted with permission. It describes the essentials of well design. While normally the authors would have read the work, paraphrased important parts, and quoted where necessary, after reading this work, it became clear that every aspect was important to share.

The work defines well integrity by one simple outcome: the prevention of vertical migration of fluids in order to protect drinking water resources. A generic shale development well is presented, beginning with its basis of design, then construction, an operational phase, and ultimately its plug and abandonment. This chronology is illustrated by a series of well schematics, which are provided in the images mentioned later as pictogram slides.

Regulations, industry standards, and best practices will be addressed, as will failure categories and relative failure rates at each phase of the well's life cycle. This case study will also raise relevant issues that may not have been fully discussed during this workshop, such as the difference between exploration and development phases, development well economics, the potential for well integrity impacts from adjacent well activities, and a time line perspective.

A brief process description for oil and gas projects might be helpful. Years before a well is drilled, significant geological and geophysical "G&G" work is performed to identify prospective areas. During this time, offset wells are studied to identify subsurface hazards that may be present in order to avoid or mitigate them. Once a prospect is defined, mineral leases are acquired, additional G&G and reservoir analysis performed, and well design determined for specific drilling locations. The first group of wells drilled are called "exploratory" and intended to define the commercial value of the prospect. Exploratory wells require extra time to gather data on the quality of the reservoir and are also used to identify well construction efficiencies for the development phase. Once the project transitions from exploration to development, each well has to pass an economic hurdle to be drilled.

Regardless of being exploratory or development, responsible oil and gas companies have a strong business incentive to protect the environment, mineral reserves, and the well itself (1). It is almost always more difficult and costly to re-enter and repair a well than addressing design deficiencies up front during construction.

This case study, although generic, is not unlike the Marcellus, Eagle Ford, and other unconventional plays with multiple hydrocarbon zones. Even though only one reservoir is the current development objective, additional reservoirs are candidates for future development. Accordingly, only the most relevant technical items such as failure modes will be included and even then, will be greatly abbreviated. For example, if corrosion is considered to be the primary failure category, the technical discussion will end there with no deeper look into the true root cause failure mode such as galvanic corrosion, sulfide stress cracking, etc. Federal and state environmental laws protect underground sources of drinking water or "USDWs." The term USDW is used synonymously with the term "protected water" and refers to an aquifer with less than 10,000 mg/l total dissolved solids or "TDS" (2). State mineral law regulates the extraction and conservation of minerals unless on federal BLM or BIA land, where federal mineral laws apply. In either case, the regulatory agency that oversees mineral extraction is also the primary regulator for protecting USDWs during oil and gas exploration and production activities (3) (4). Protected water and hydrocarbons have natural separation (5) in most situations. There are however, areas of the country where methane is routinely found to exist naturally in USDWs (6) (7) and has been associated with bubbles in rivers as early as the mid-1800s (8). There are also locations where methane vents to the surface via natural pathways having nothing to do with oil and gas extraction activities (9) (10). It has been estimated from a review of Pennsylvania regulatory records that over 95% of the complaints that oil and gas activities had contaminated private water wells were actually due to preexisting or other land use activities (11). These naturally occurring migrations are not limited to methane, as towns named Oil Springs, KY (12), Oil

Springs, Ontario (13), and historical sites such as Seneca Oil Spring, NY (14) and Brine Springs, TX (15) all attest that oil and brine have been observed migrating to the surface dating back to the 1600s.

5.1 Basis of Design

A development well is drilled only if there is confidence that the estimated recoverable hydrocarbon reserves will provide an acceptable economic rate of return, given the cost to construct and operate the well. For an unconventional gas play, development wells tend to have generational designs where a group of wells will have a similar drilling, casing, cementing, perforating, and hydraulic fracturing design. Over time as more wells are drilled, experience provides opportunities to correct any design deficiencies, improve drilling efficiencies, and well performance, and therefore subsequent generations of wells are seldom designed exactly the same.

Individual wells, regardless of their generational status, receive detailed engineering analysis and planning that is communicated to the well site supervisor in the form of a written drilling and completion procedure. These well-specific procedures are a planned sequence of activities that also incorporate regulatory compliance and industry best practices.

6

Well Construction

6.1 Drilling

A typical onshore well is a conductor pipe that is driven, drilled, or augered into the ground by a construction crew or "spud rig" prior to the drilling rig's arrival. This conductor pipe is a structural component that sometimes is not needed at all. Conductor pipe most often does not reach the top and does not penetrate the base of protected water; therefore, it is not involved in protecting USDWs from vertical migration of fluids. Accordingly, failure categories for the conductor pipe will not be discussed.

The surface hole is drilled to a prescribed depth below the base of protected water. This depth is most often provided by the State Oil and Gas Regulator, as in Oklahoma (16), or the State Environmental Protection Regulator, as in Texas (17), or not specifically provided other than to protect all USDWs encountered, as in Pennsylvania (18). In this latter situation, oil and gas operators typically research a Pennsylvania Groundwater Information System "PaGWIS" database and local water well driller's records to generate a hydrogeological map in order to determine depths of water that need to be protected.

The surface hole is not left open for more than a few hours while being drilled, cased, and then cemented back to surface. Those zones left open during this brief period are all USDWs, so vertical migration of fluids does not present a significant threat during surface hole drilling. The surface hole on our case study well is drilled in a few hours on day #1of the drilling operation.

The surface casing string is the primary barrier to prevent fluids from the wellbore from entering protected water as the well is being drilled to the next casing setting depth. Unlike the conductor pipe, surface casing is always required, and is typically specified by regulation to be of "suitable and sufficient" quality (19) or "suitable for all drilling and operating conditions such as tension, burst, collapse" (20). For all casing strings, industry best practices provide extensive guidance on the selection of proper casing size, grade, weight, and connections, plus procedures for field handling, inspection, and testing (21) (22) (23) (24) (25) (26) (27). For our case study well, the surface casing is "run" or installed in a few hours during day #1 of the drilling operation.

Failure categories for the surface casing and all other casing strings can be divided into the following five categories (28). It should be noted that two of these categories, mechanical and corrosion, may be secondary to cement failures where a failed cement sheath can lead to buckling or external corrosion that would not have otherwise occurred. Failure categories, their respective failure modes, relative failure rates, and remedial options will be discussed briefly:

- Materials: defects, tolerance busts, not getting the quality of pipe specified
- Connections: wrong connection selected for the service, improper makeup
- Wear and Handling: internal wear from drilling, external damage from handling
- Mechanical: tensile, burst, collapse, buckling, cyclic loading
- Corrosion: internal vs external; galvanic, CO2, sulfide stress, hydrogen induced cracking

Materials defects are supplier dependent, and can be managed by inspections and other supply chain quality control efforts. Connection problems are most often related to improper makeup and can be minimized by onsite supervision. Wear for the surface casing string is seldom a concern and occurs as a result of other problems encountered while drilling the well. Mechanical problems with the surface casing are very few when compared to deeper casing strings that are exposed to higher pressures and temperatures. External corrosion presents the highest failure category for surface casing. Remedies may include external coatings, cement squeezes, and cathodic protection systems.

The surface casing string's cement job provides the primary barrier against vertical migration of fluids into protected water for the entire life of the well. In the context of USDW protection, the importance of getting a good primary cement job on the surface casing string cannot be overstated. Remedial cementing options do not provide high success rates for zonal isolation and should be considered only for contingency purposes. Of all regulations for onshore wells, the rules for surface casing cementing contain the most stringent requirements for hole size versus casing size, centralization, cement quality, cement quantity, cement placement techniques, and quality assurance than for any other casing string (29). Failure to properly cement the surface casing string triggers both agency notification and corrective actions (30). The surface casing on our case study well is cemented on day #2 of the drilling operation.

There is a significant body of information published on cement selection and cementing best practices (31) (32) (33) (34). There is also a significant body of information available on cementing failure rates (35) (36). This Well Integrity Case Study will focus on those conditions that directly relate to zonal isolation for the protection of USDWs, briefly discussing three failure categories, with their respective modes and relative failure rates, and remedial options:

- Insufficient cement volume: underestimated annular volume, lost circulation
- Low bond strength: poor slurry design, poor management of hydrostatic head pressure
- Micro annulus, cracking, plastic deformation: thermal and pressure effects, cyclic loads

Cement failure rates are directly proportional to the ability to evaluate the top of and quality of the cement sheath. Cement tops can be identified by a temperature log, relative cement bond quality can be identified by a Cement Bond Log or CBL, while absolute cement bond quality requires a combination of logging, testing, and engineering analysis (37).

For all three cement failure categories, remedial options are not optimum, and include pumping in from the top, spotting from the top via a small work string, or by perforating and squeezing. It should be noted that two of these three remedies, pumping in from the top and perforating and squeezing, might add new problems for zonal isolation if not properly executed.

There is a strong correlation between gas migration and uncemented or poorly cemented casing strings. There is also a strong correlation between external casing corrosion and the absence of a good cement sheath (35) (36). After the surface casing has been successfully tested, the float collar, float shoe, and approximately 10 feet of new formation are drilled. Another integrity test is then performed, a Formation Integrity Test or "FIT," which tests both the casing shoe and new formation together. This is not a leak-off test and does not test the limits of the shoe and formation; rather, the FIT provides an assessment of the wellbore's ability to withstand additional pressure in case of an influx of fluids and allows for safer drilling to the next casing point (38).

The next sections of well, which for this case study includes an intermediate and production casing section, are essentially a repeat of the surface casing section described above, except that:

- The design depth for intermediate and production casing strings are not as comprehensively regulated (as for the surface casing depth) other than to provide safe drilling operations and to prevent the waste of minerals.
- The regulations concerning hole size versus casing size, centralization, cement quality, cement quantity,

> cement placement techniques, and quality assurance for intermediate and production casing strings are not as specific (as for the surface casing) other than to provide safe drilling operations and prevent the waste of minerals.

Although this case study well has been drilled, cased, and cemented over a 30 day period, the first two days are the most critical for zonal isolation of USDWs where the foundation for well integrity is determined.

6.2 Completion

Well completion is the where the production casing is perforated, the formation is hydraulically fractured, frack fluids are unloaded from the formation, and production operations commence. This is basically the well's configuration for the rest of its life as it relates to protecting USDWs.

Prior to performing the hydraulic frack, the production casing is tested to anticipated frack pressure plus a safety factor, as is the frack tree and all of the surface pumping equipment and lines. During the frack, all casing annuli are monitored, as are the injection rate, injection pressure, and slurry properties. If during the frack job, significant pressure is found on the intermediate casing annulus, or there is any indication of communication with the surface casing annulus, the frack job is shut down and not resumed until corrective actions are made that only the intended zone is subject to frack pressures.

Refracks are similar to original fracks as discussed above, with the exception that a frack string or wellhead saver might be used to protect older production casing strings and wellheads from frack pressures. This is a case-by-case situation that requires additional testing and engineering analysis in order to protect both the well and USDWs during refrack operations.

As the well is produced, reservoir pressures tend to drop and liquid rates tend to rise; therefore, devices for lifting liquids, such as a tubing string with pumping or gas lift equipment, become necessary. This internal configuration can have an impact on USDW protection and is addressed during the operations phase.

Below is a schematic of the various layers in a typical site casing.

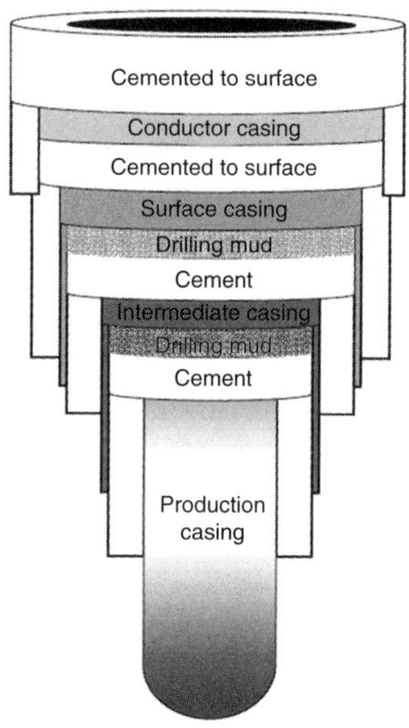

Figure 6.1 Well Casing.

7

Well Operations

Prudent operators monitor all casing annuli on a regular basis to be able to detect sustained casing pressure or SCP. This condition could be caused by thermal expansion of annular fluids, packer or liner leaks, leaks into the annulus from inner tubing or casing strings, or from annular migration due to poor zonal isolation.

All states have rules for reporting and responding to the loss of well integrity, which includes releases, non-thermal SCP, and other abnormal situations (39), as does the BLM (40), as do best industry practices (41). The Commonwealth of Pennsylvania has new rules that require quarterly mechanical integrity testing and annual reporting for all operating wells (42).

Adjacent well operations may have an impact on mechanical integrity as our case study well contemplates. Hydraulic fracturing of a well near our case study well into a zone that is not protected, or not adequately protected for the conditions imposed, can lead to unwanted well to well communication. This is currently a void where regulations and industry practices have not fully recognized that well integrity can become a neighborhood issue.

7.1 Well Plug and Abandonment "P&A"

Similar to well construction regulations and industry practices, well P&A also has comprehensive guidance to prevent vertical migration of fluids into USDWs. There is clear guidance for plug location, cement quantity, quality, placement techniques, testing, and reporting (43) (44) (45). Regulations may also specify that only approved cementing contractors perform plugging, require independent onsite supervision, and require post cement job certifications by both the operator and the cementing company. There are also significant industry studies and best practices for well P&A (46) (47).

Failure studies have found that vertical migration issues in P&A wells are directly related to the original primary cement job during well construction. Those wells with gas migration to the surface prior to well P&A were likely to continue to have gas migration to the surface after P&A. Additionally, those wells plugged with bridge plugs and dump bailed cement on top were found to be more prone to leakage than wells plugged with cement that was circulated or squeezed in place (35) (36).

7.2 Considerations

Well integrity and well construction are inextricably linked, regardless of the completion technique selected. Primary cementing is the critical step for preventing vertical migration of fluids during the well's productive life, and afterwards. State and federal regulations address casing and cementing with prescriptive rules and reporting requirements, while industry employs a large body of technical studies and best practices. There are five identified casing failure categories: materials, connections, wear/handling, mechanical, and corrosion. These are not as problematic for zonal isolation as three identified cementing failure categories: insufficient cement volume, low bond strength, and cement sheath damage.

For hydraulically fractured completions, significant bodies of industry technical information and best practices have been published. State and federal regulations address hydraulic fracturing with rules and reporting requirements that are continuously adapting to keep pace with technology advancements (48). Adjacent wells and the potential for unwanted communication during hydraulic fracturing is a concern. State and federal regulations are largely silent on this issue, as are industry studies and best practices.

EPA Hydraulic Fracturing Study Technical Workshop #2 March 10-11, 2011
Well Integrity Case Study by Lloyd H. Hetrick
Slide 1 of 22 – Well Integrity Narrowly Defined as the Prevention of Fluids Migration into Protected Water

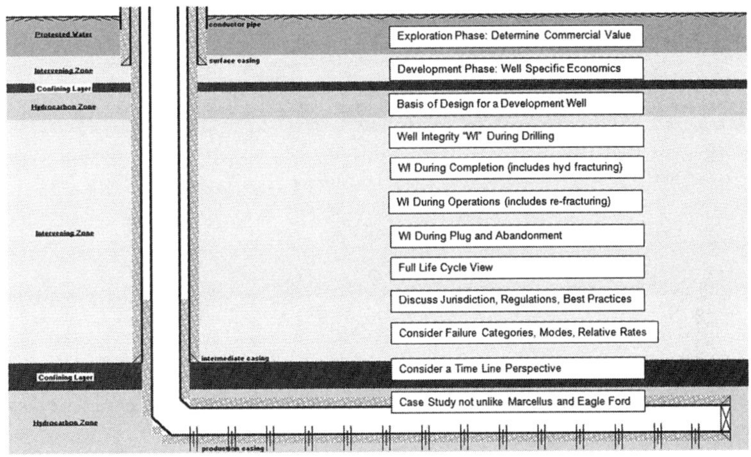

EPA Hydraulic Fracturing Study Technical Workshop #2 March 10-11, 2011
Well Integrity Case Study by Lloyd H. Hetrick
Slide 2 of 22 – Jurisdiction

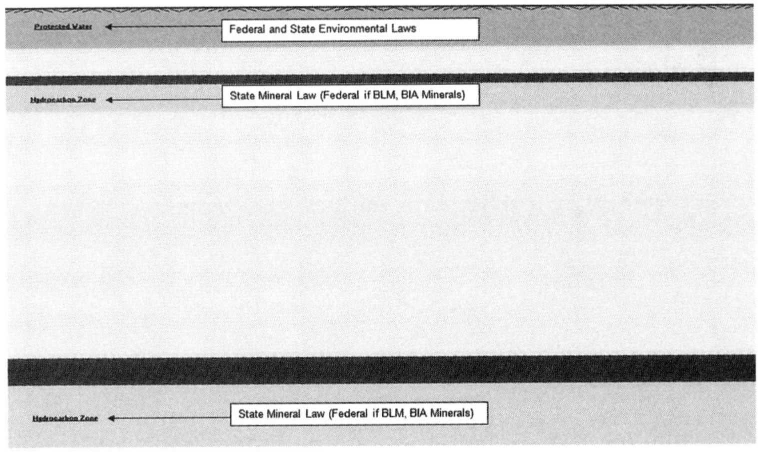

EPA Hydraulic Fracturing Study Technical Workshop #2 March 10-11, 2011
Well Integrity Case Study by Lloyd H. Hetrick
Slide 3 of 22 – Natural Separation of Protected Water and Hydrocarbons, in most cases

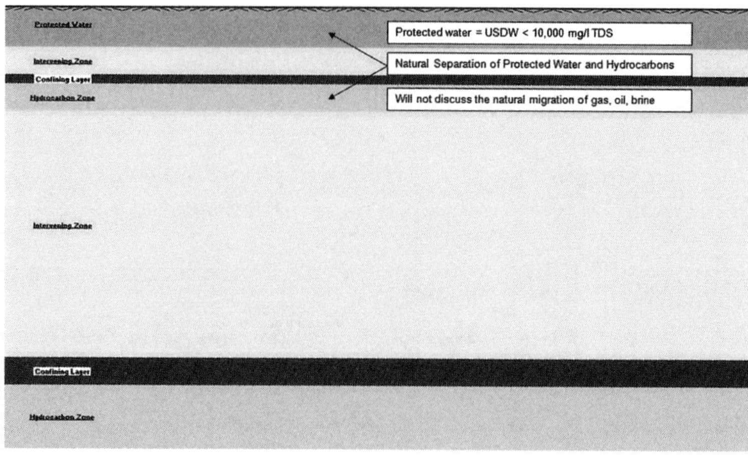

EPA Hydraulic Fracturing Study Technical Workshop #2 March 10-11, 2011
Well Integrity Case Study by Lloyd H. Hetrick
Slide 4 of 22 – Spud well with conductor pipe (driven, augered or drilled) no impact on Protected Water

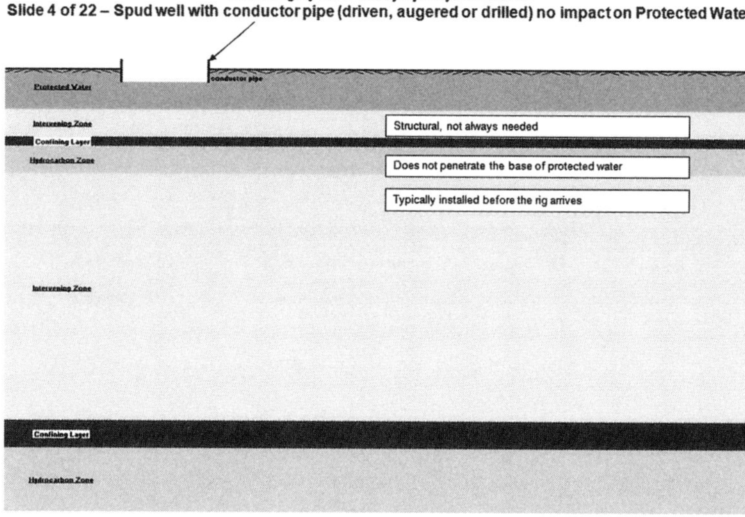

EPA Hydraulic Fracturing Study Technical Workshop #2 March 10-11, 2011
Well Integrity Case Study by Lloyd H. Hetrick
Slide 5 of 22 – Well is drilled to a prescribed depth below Protected Water

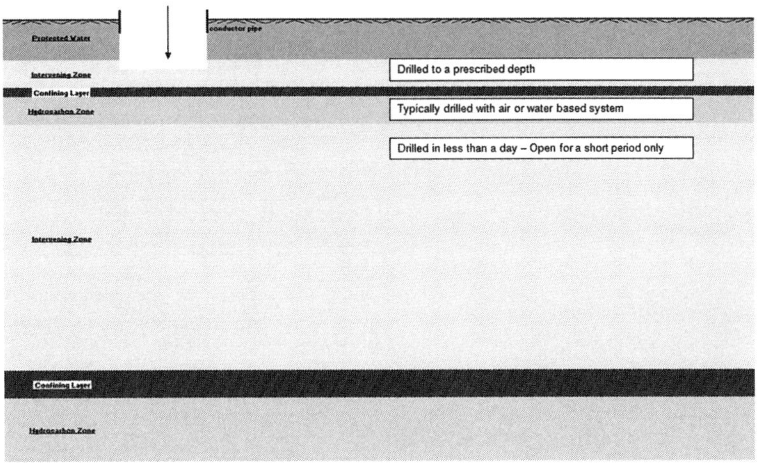

EPA Hydraulic Fracturing Study Technical Workshop #2 March 10-11, 2011
Well Integrity Case Study by Lloyd H. Hetrick
Slide 6 of 22 – Surface casing is run

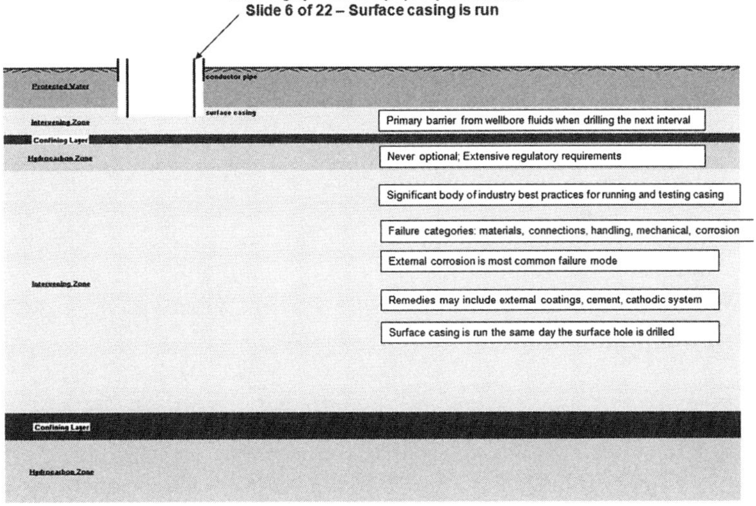

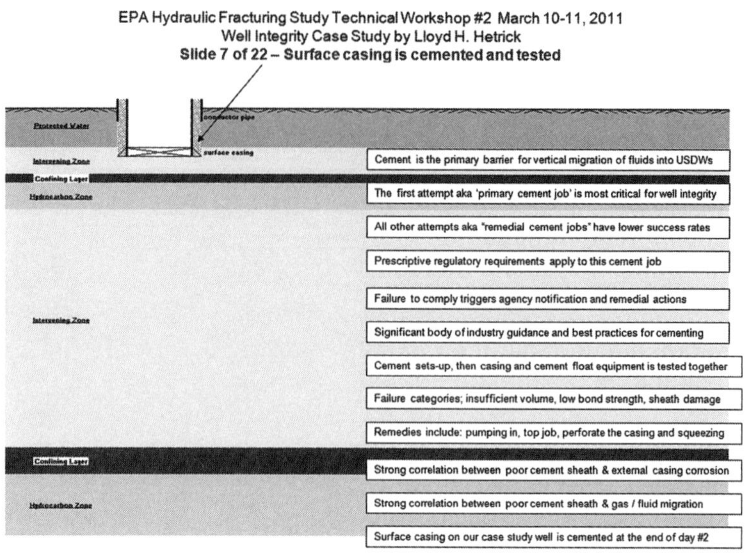
EPA Hydraulic Fracturing Study Technical Workshop #2 March 10-11, 2011
Well Integrity Case Study by Lloyd H. Hetrick
Slide 7 of 22 – Surface casing is cemented and tested
Protected Water
conductor pipe
surface casing
Intervening Zone
Confining Layer
Hydrocarbon Zone
Intervening Zone
Confining Layer
Hydrocarbon Zone
Cement is the primary barrier for vertical migration of fluids into USDWs
The first attempt aka 'primary cement job' is most critical for well integrity
All other attempts aka "remedial cement jobs" have lower success rates
Prescriptive regulatory requirements apply to this cement job
Failure to comply triggers agency notification and remedial actions
Significant body of industry guidance and best practices for cementing
Cement sets-up, then casing and cement float equipment is tested together
Failure categories; insufficient volume, low bond strength, sheath damage
Remedies include: pumping in, top job, perforate the casing and squeezing
Strong correlation between poor cement sheath & external casing corrosion
Strong correlation between poor cement sheath & gas / fluid migration
Surface casing on our case study well is cemented at the end of day #2

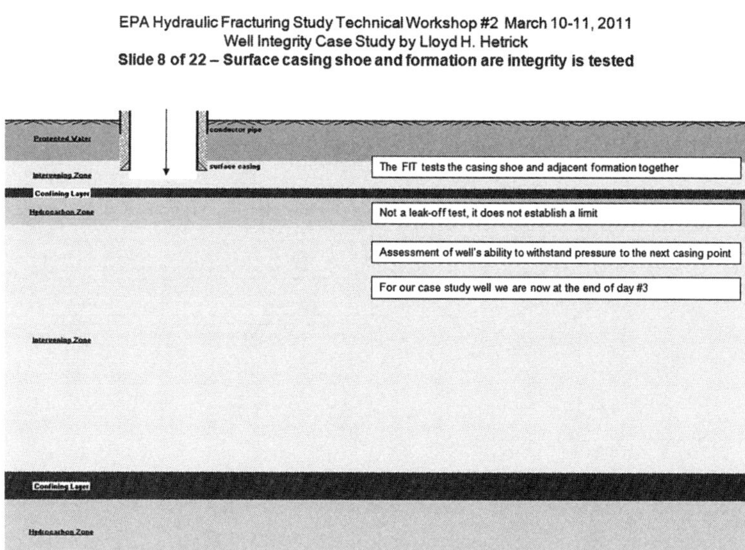
EPA Hydraulic Fracturing Study Technical Workshop #2 March 10-11, 2011
Well Integrity Case Study by Lloyd H. Hetrick
Slide 8 of 22 – Surface casing shoe and formation are integrity is tested
Protected Water
conductor pipe
surface casing
Intervening Zone
Confining Layer
Hydrocarbon Zone
The FIT tests the casing shoe and adjacent formation together
Not a leak-off test, it does not establish a limit
Assessment of well's ability to withstand pressure to the next casing point
For our case study well we are now at the end of day #3
Intervening Zone
Confining Layer
Hydrocarbon Zone

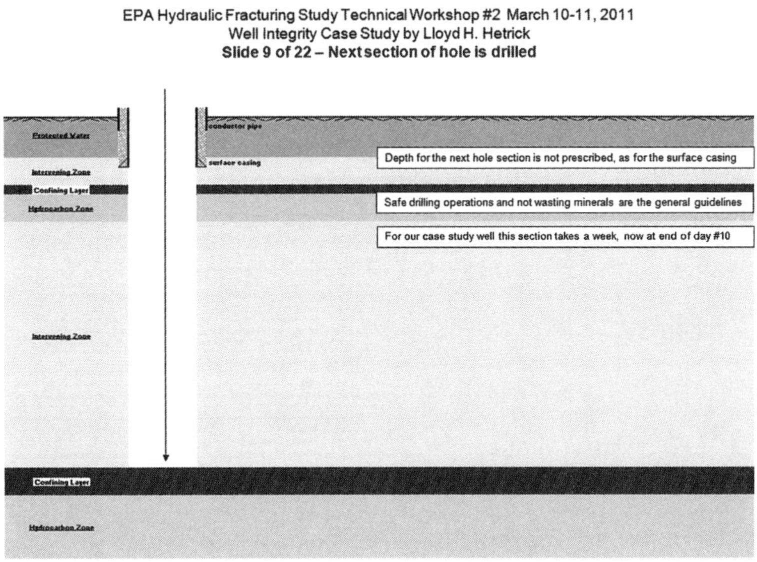
EPA Hydraulic Fracturing Study Technical Workshop #2 March 10-11, 2011
Well Integrity Case Study by Lloyd H. Hetrick
Slide 9 of 22 – Next section of hole is drilled
Protected Water
conductor pipe
Intervening Zone
surface casing
Depth for the next hole section is not prescribed, as for the surface casing
Confining Layer
Hydrocarbon Zone
Safe drilling operations and not wasting minerals are the general guidelines
For our case study well this section takes a week, now at end of day #10
Intervening Zone
Confining Layer
Hydrocarbon Zone

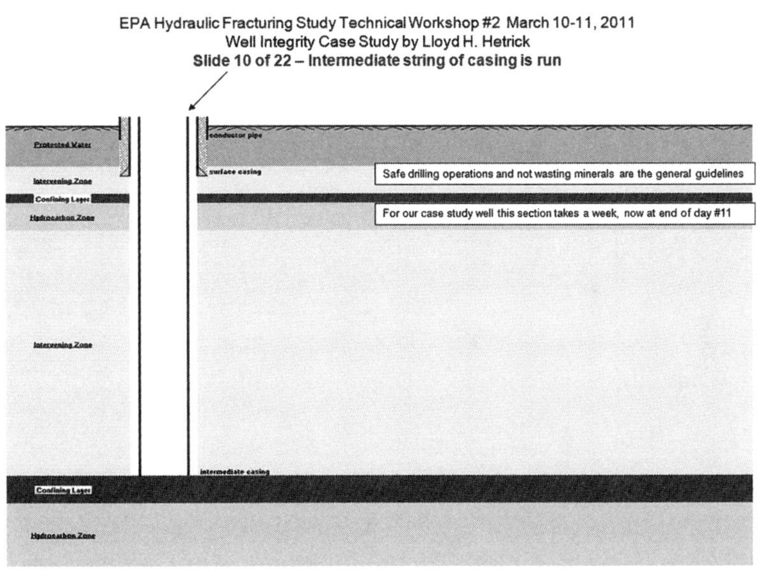
EPA Hydraulic Fracturing Study Technical Workshop #2 March 10-11, 2011
Well Integrity Case Study by Lloyd H. Hetrick
Slide 10 of 22 – Intermediate string of casing is run
Protected Water
conductor pipe
Intervening Zone
surface casing
Safe drilling operations and not wasting minerals are the general guidelines
Confining Layer
Hydrocarbon Zone
For our case study well this section takes a week, now at end of day #11
Intervening Zone
intermediate casing
Confining Layer
Hydrocarbon Zone

EPA Hydraulic Fracturing Study Technical Workshop #2 March 10-11, 2011
Well Integrity Case Study by Lloyd H. Hetrick
Slide 11 of 22 – Intermediate casing is cemented and tested

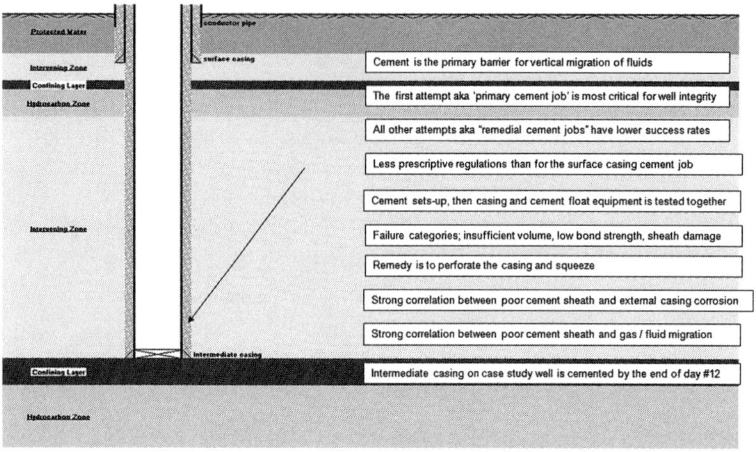

EPA Hydraulic Fracturing Study Technical Workshop #2 March 10-11, 2011
Well Integrity Case Study by Lloyd H. Hetrick
Slide 12 of 22 – Intermediate casing shoe and formation integrity are tested

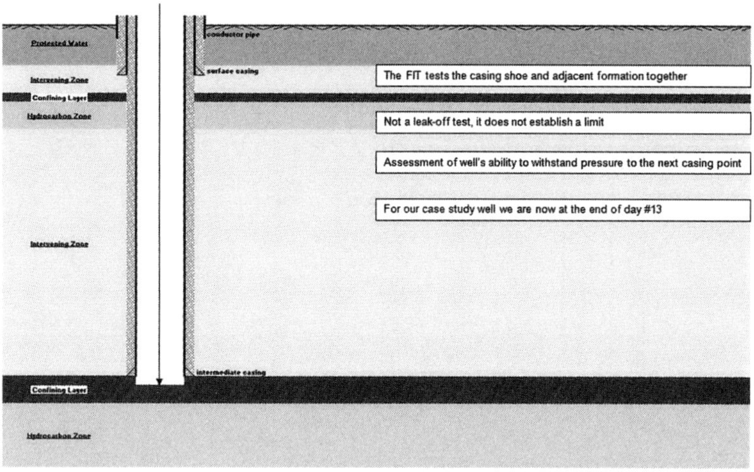

EPA Hydraulic Fracturing Study Technical Workshop #2 March 10-11, 2011
Well Integrity Case Study by Lloyd H. Hetrick
Slide 13 of 22 – Next section of hole is drilled

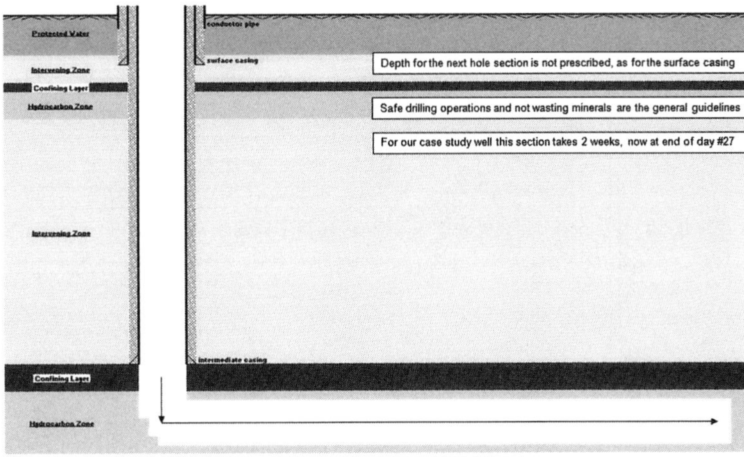

EPA Hydraulic Fracturing Study Technical Workshop #2 March 10-11, 2011
Well Integrity Case Study by Lloyd H. Hetrick
Slide 14 of 22 – Production casing is run

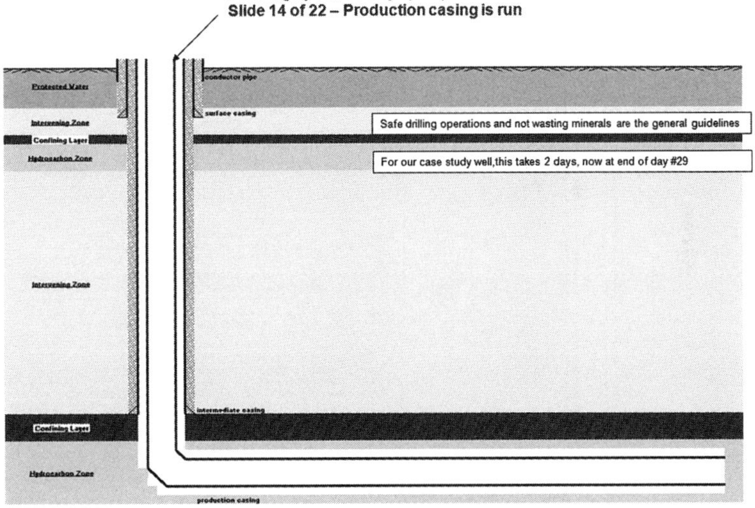

EPA Hydraulic Fracturing Study Technical Workshop #2 March 10-11, 2011
Well Integrity Case Study by Lloyd H. Hetrick
Slide 15 of 22 – Production casing is cemented, Drilling phase is done

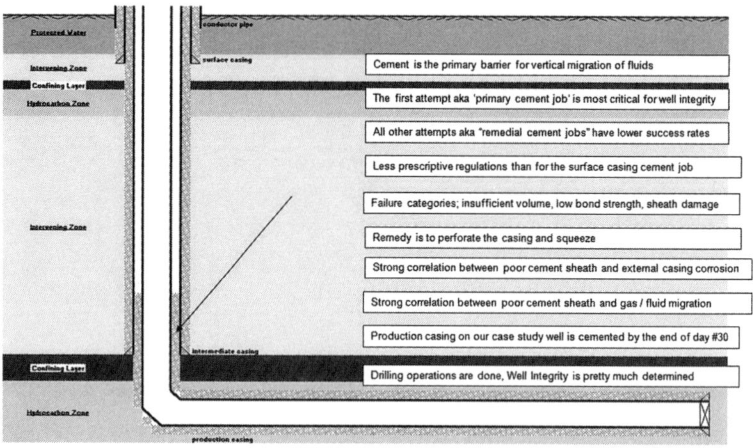

EPA Hydraulic Fracturing Study Technical Workshop #2 March 10-11, 2011
Well Integrity Case Study by Lloyd H. Hetrick
Slide 16 of 22 – Completion phase: perforate, hydraulically frac, flowback, production begins

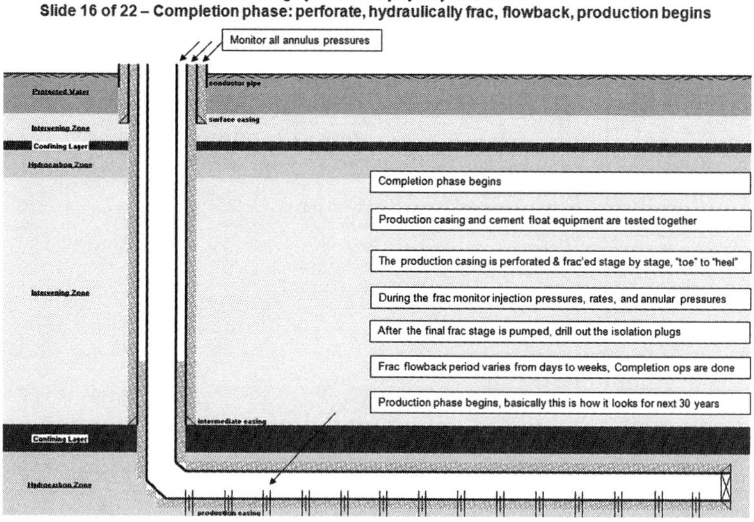

EPA Hydraulic Fracturing Study Technical Workshop #2 March 10-11, 2011
Well Integrity Case Study by Lloyd H. Hetrick
Slide 17 of 22 – Refrac and Adjacent Well Frac Scenarios

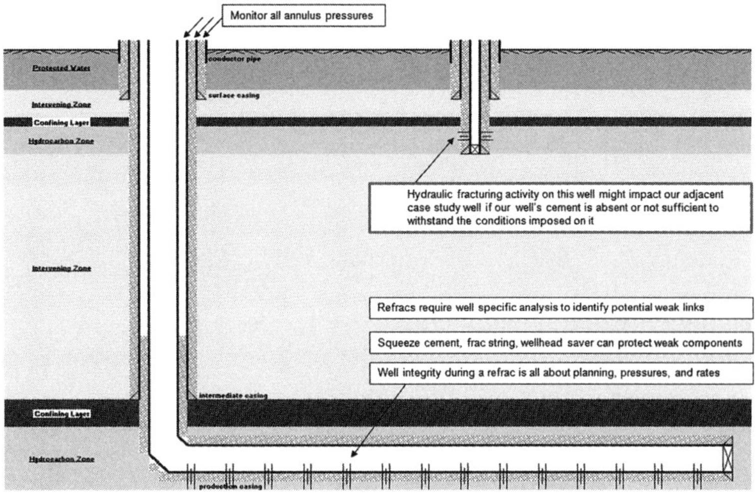

EPA Hydraulic Fracturing Study Technical Workshop #2 March 10-11, 2011
Well Integrity Case Study by Lloyd H. Hetrick
Slide 18 of 22 – Plug and Abandonment begins, producing zone is plugged and tested

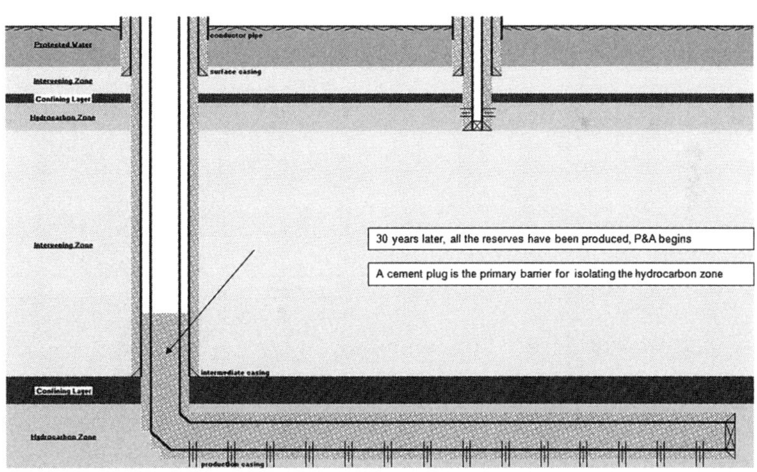

EPA Hydraulic Fracturing Study Technical Workshop #2 March 10-11, 2011
Well Integrity Case Study by Lloyd H. Hetrick
Slide 19 of 22 – Casing is cut or perforated near the surface casing shoe

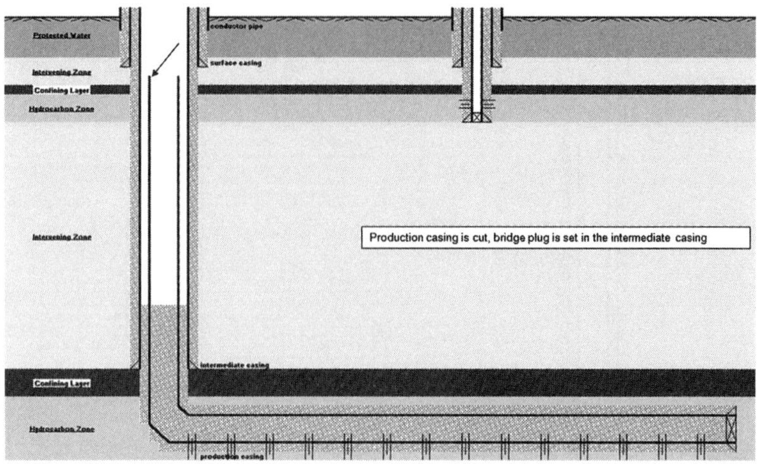

EPA Hydraulic Fracturing Study Technical Workshop #2 March 10-11, 2011
Well Integrity Case Study by Lloyd H. Hetrick
Slide 20 of 22 – Cement plug is set across the surface casing shoe and tested

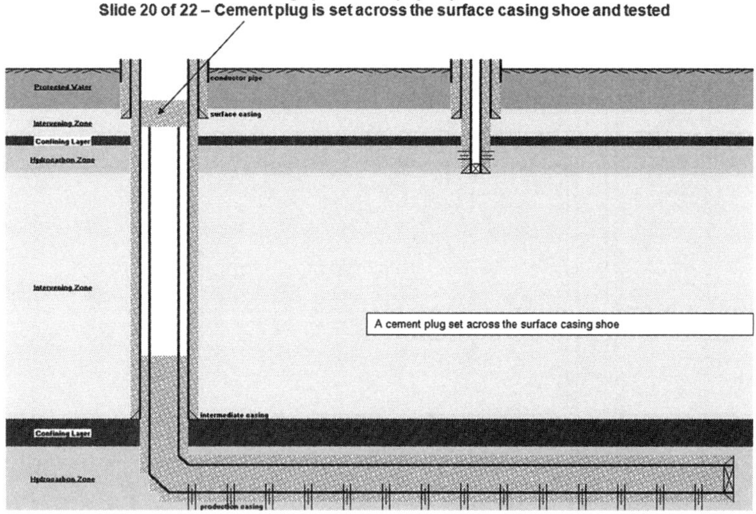

EPA Hydraulic Fracturing Study Technical Workshop #2 March 10-11, 2011
Well Integrity Case Study by Lloyd H. Hetrick
Slide 21 of 22 – Shallow surface plug is set and tested

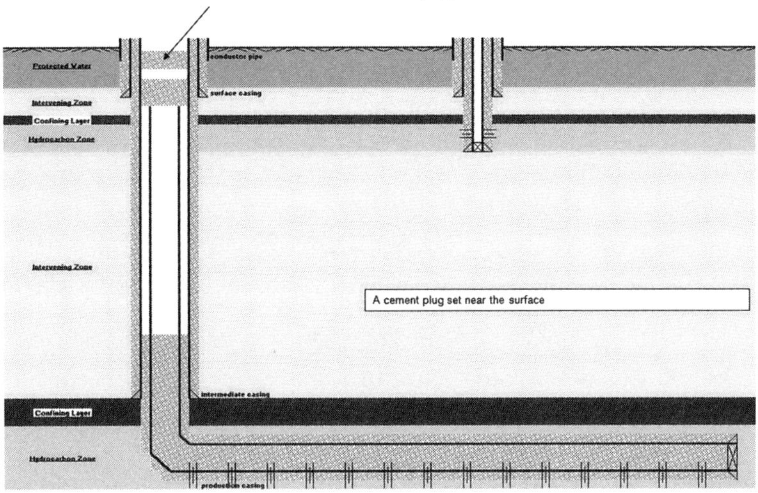

EPA Hydraulic Fracturing Study Technical Workshop #2 March 10-11, 2011
Well Integrity Case Study by Lloyd H. Hetrick
Slide 22 of 22 – Pipe cut-off, cap welded, well ID marker, backfill

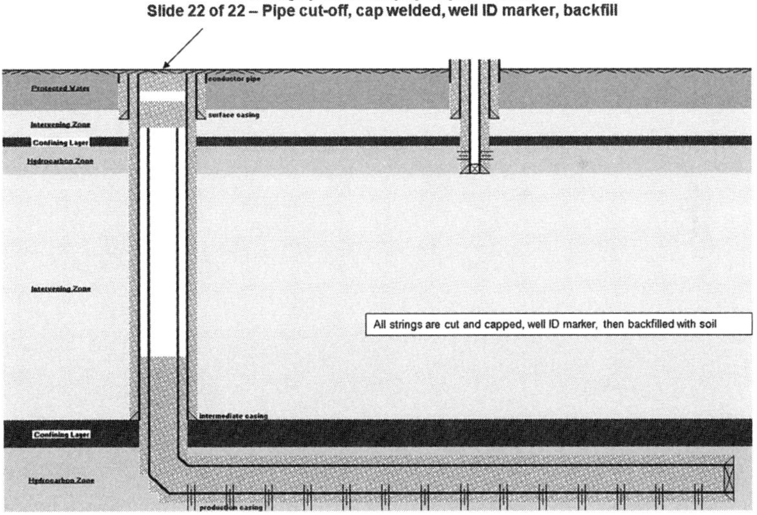

8

Failure and Contamination Reduction

The following are several key areas to be addressed in order to develop a pad site that is both safe for the environment as well as being productive.

8.1 Conduct Environmental Sampling Before and During Operations

Many of the contaminants that have been attributed to fracking, such as methane gas, occur naturally in groundwater and soil. If post-fracking contamination of soil or water is found, and there is no pre-fracking baseline against which the contamination can be compared, it could be difficult for the well operator to demonstrate that such contamination was pre-existing. This type of uncertainty increases the risk of regulatory scrutiny and litigation. Some states, like Wyoming, already require pre-drilling disclosures. Consider engaging a qualified environmental consultant to test water and soil before fracking begins. Air modeling or monitoring should also be performed during operations to help demonstrate that emissions are under control.

8.2 Disclose the Chemicals Being Used in Fracking Operations

This may be the issue that generates the most attention from state legislators and environmentalists. All of the states that recently have enacted fracking related statutes (e.g., Texas, Colorado, Ohio, Pennsylvania) have included sections that not only address the disclosure of chemicals used, but also provide some level of protection for trade secrets. The energy industry as a whole has supported this approach and generally agrees that while the disclosure of the chemicals used is appropriate, companies should not be forced to publish their exact fracking fluid "recipes." Companies have also come to recognize that there are benefits to using environmentally "friendly" chemicals as much as possible. For example, when the public is advised that the guar gum used to gel frack water is the same ingredient commonly found in ice cream and bakery products, this type of disclosure tends to negate the fear associated with the use of "unknown" chemicals.

8.3 Ensure that Wellbore Casings are Properly Designed and Constructed

The U.S. Department of Energy demonstrated that the risks of a properly constructed well contaminating an aquifer as a result of racking are remote. The most significant risks come instead from poorly constructed casings that might leak into groundwater. Thus, many state legislatures have enacted strict regulations that govern the manner in which casings are to be constructed, tested and monitored. Once again, the industry as a whole has already embraced these efforts to ensure that wells are properly built and should continue to implement improved technologies as they become available.

8.4 Eliminate Venting and Work Toward Green Completions

On April 17, 2012, the U.S. Environmental Protection Agency (EPA) issued the first federal air standards for natural gas wells that are hydraulically fractured. A key component of the final rules is the implementation of a process known as "green completion." The process separates gas and liquid hydrocarbons from the flowback and is expected to yield a nearly 95 percent reduction in greenhouse gases. Although green completions are not mandatory until January of 2015, companies are expected to use this time to acquire the necessary equipment. Until then, flaring is to be used to minimize the impact of these gases upon the environment.

8.5 Prevent Flowback Spillage/Leaks

Millions of gallons of flowback can be generated by fracking just one well. The flowback will include sand, chemicals (i.e., biocides, surfactants, gelling agents), brine, and dissolved solids. If the flowback is not contained properly, it can leak into nearby surface water and soil, thereby harming the environment and increasing the risk for various types of litigation and regulatory action. One way to minimize such risks is to ensure that all flowback pits are properly lined. Another is to implement a "closed system" that eliminates the pit altogether and routes the flowback directly to storage tanks prior to offsite treatment and disposal.

8.6 Dispose/Recycle Flowback Properly

The EPA and states share responsibility for implementing the Clean Water Act (CWA) programs. The effluent guidelines program prohibits the on-site direct discharge of flowback from fracking wells into waters of the U.S. While some flowback is transported to publicly owned treatment works (POTWs), and significant amounts are still injected into underground wells for disposal, there is an increasing trend to recycle the flowback, and for good reason. Recycling can significantly reduce the amount of water needed for fracking activities – an extremely important consideration in those portions of the country experiencing drought conditions. And if the recycling is done near the drilling pad, it can also curtail the truck traffic needed to deliver the water and remove the flowback, thereby lowering transportation and offsite disposal costs.

8.7 Minimize Noise and Dust

Fracking activities at the drill pad can be loud. When they occur in urban areas or near residential communities, efforts are frequently undertaken to abate the noise by surrounding the drill pad with noise blankets, sound curtains, or wall barrier systems. Truck traffic also has the potential to generate significant levels of noise and dust. As indicated above, reducing the amount of traffic at the drill pad by recycling is one way to combat these potential problems. Other ways include the use of centralized pump stations and multi-well drilling techniques that reduce the number and size of pads needed for fracking. Spraying service roads with treated production water and brine is also permitted in some states to reduce dust, but care must be taken not to apply such liquids near vegetation, residences, or drainage ditches.

8.8 Protect Workers and Drivers

The New York Times has reported that the most dangerous jobs associated with fracking are not at the drilling rig but on the road. Drivers working at oil and natural gas wells reportedly work longer hours than drivers in other industries, thereby subjecting them to fatigue and a greater risk of having an accident. That is not meant to minimize the risks faced by drilling crews who obviously could be exposed to a many potential jobsite hazards if proper safety precautions are not followed. While certain exemptions for oil and gas truckers were upheld by the Federal Motor Carrier Safety Administration, the exemptions have come under closer scrutiny in the last two years. The need for companies to enforce OSHA regulations and other safety rules implemented by the industry to protect its workers is of critical importance.

8.9 Communicate and Engage

Nearly every group that has generated a set of Best Management Practices has emphasized the importance of communicating with nearby residents, local governments, and other stakeholders. Misinformation can be quickly spread through the Internet. First impressions, although mistaken, can be difficult to reverse. The Investor Environmental Health Network (IEHN) recommends that the fracking industry identify communities potentially impacted, address their major concerns, and establish a mechanism for resolving conflicts. Companies should be prepared to direct persons with legitimate questions to websites that provide answers that are scientifically and factually accurate. Several such websites exist, such as http://www.energyindepth.org/. The need for patience must also be stressed, since many of the discussions that involve fracking are emotionally charged.

8.10 Record and Document

This last recommendation comes from the authors who also happen to be litigators. Sloppy record keeping gives plaintiffs, regulators, and juries too much leeway to draw erroneous conclusions. If documentation should have been kept, but was not, there will almost always be an inference that the "lost" documents were adverse to the interests of the company. Sloppy record keeping reflects poorly on a company's overall reputation even though the company may have an excellent history of actual environmental compliance and worker safety. It is also important to recognize that in this fast-paced age of instant communications with emails and text messages, virtually all of those communications can be converted to documents that must be produced if litigation occurs. Thus, to the extent possible, refrain from sending emails and text messages when frustrated or angry. Similarly, jokes and needless exaggeration usually do not translate well. Think about how that email might look if blown up on a screen in front of a jury. If that image causes discomfort, think twice before sending it.

9

Frack Fluids and Composition

The use of hydraulic fracturing for oil and gas exploration in the U.S. has become highly controversial, with one of the greatest points of contention between the public and industry being the make-up of frack fluids and their possible impacts on public health and the environment. This has become such a hot topic with many segments of the public for two reasons:

1. if a concern exists about the pumping of fluids into any structure, then the most concern will naturally be centered on what is being pumped, and
2. a great deal of suspicion arose and was intensified when the oil and gas industry initially balked at the disclosing the chemical makeup of fluids used to enhance hydraulic fracturing.

This has become a major argument point for the concerned public because, basically, "if there is nothing to hide, why not disclose it?" Take, for example, if a small child walks into a room with his hands behind his back and will not show what is there for several minutes...and then only does when forced. Well, even if it turns out he was holding something as harmless as a feather behind his back it will cause suspicion all the same, wondering "what was he doing with that feather!"

To make this contentious subject a little clearer, this section will provide descriptions of why frack fluids are needed, what general chemicals are needed and used, relative amounts of chemicals in frack fluid composition, the different types and uses of proppants, a discussion on slickwater, and a discussion on present regulations and standards for industry disclosure of frack fluid compositions.

9.1 Uses and Needs for Frack Fluids

There are a great deal of varied chemicals used every day in oil production wells during all phases of drilling, completions, and production. These chemicals can include cements used to seal the annulus to protect the pipe and surrounding formation from damage through wells exceeding the producing and stimulation requirements placed on the pipe, temperature, and even natural ground stresses. An example is corrosion inhibitors. These chemicals help pipe and connection seals remain within design specifications to prevent failures. Corrosion prevention and treating chemicals may also be necessary due to operational and field changes, even after well completion and during production.

These chemicals can be used in much the same fashion as fracturing; however, chemicals in well operation are applied in smaller quantities, at lower pressure, and in a regular maintenance-driven schedule during a well's life. Just like the maintenance-driven chemicals utilized during operations, chemicals serve numerous necessary functions to ensure successful, safe, and efficient hydraulic fracturing operations. The following provides a comprehensive look at common chemical additives utilized in the current fracturing industry.

9.2 Common Fracturing Additives

There is no one formula for how much each of the following additives are used in a given fracturing fluid; however, the following section is intended to present a brief description of some of the most commonly used additives and a general percentage breakdown of each that has been widely reported. Each well differs in the number, type, and amount of additives used (please note: the term "additives" is used to include water, sand, and chemicals to allow for a discussion of each under one heading) in a successful fracture treatment; "typically," between 3 and 12 additives are used, depending on the conditions of the specific well to be fractured and characteristics of the surrounding formations. Additives utilized in hydraulic fracturing operations are intended to serve specifically engineered uses, such as biocides to control microorganism/bacterial growth, corrosion inhibitor to prevent corrosion of pipe, viscosity agents to carry proppant, gelling agents to improve proppant placement, friction reduction to decrease pump friction and reduce treating pressure, oxygen scavengers to also aid in corrosion prevention in metal pipes, and acids to help remove drilling mud build up damage.

Fluids (typically water): usually approximately 98%-99% of the total volume; used to create the fractures in the formation and to carry a propping agent (typically silica sand) that is deposited in the induced hydraulic fractures to keep them from closing up.

Hydrochloric acid (HCl): from 5% to 25% in solution used to help dissolve minerals and help remove damage near the well bore by cleaning out cement around pipe perforations; also helps initiate fissures in the rock matrix.

Corrosion inhibitor (typically ammonium bisulfate): usually approximately 0.2%-0.5% of acid total volume, resulting in approximately 5–10 gallons; used only in instances when acid is used to prevent pipe corrosion.

Biocides (typically sodium hypochlorite or chlorine dioxide): usually approximately 0.005%-0.05% of the total volume; used to control bacterial growth in the water injected into the well and prevent pipe corrosion.

Friction reducers (typically polyacrylamide based compounds): usually approximately 0.025% of total volume; used to reduce pipe friction and pressure in the piping required to pump fluids.

Gelling agents (guar gum and cellulose): not often used; used to thicken water-based solutions and help in suspension and transport of proppants into formation.

Crosslinking agent (boric acid, titanate and zirconium): used to enhance abilities of the gelling agent to even further aid in transport of proppant material.

Breaker solution: when cross-linking additives are added, a breaker solution is commonly added in the frack stage to cause the enhanced gelling agent to break down into a simpler fluid so it can be readily removed from the wellbore without carrying back the sand/ proppant material.

Oxygen scavenger (ammonium bisulfate): used to prevent corrosion of pipe by oxygen.

Iron control and stabilizing agents (citric acid and acetic acid): used to keep iron compounds in soluble form to prevent precipitation.

Surfactant: usually approximately 0.5 to 2 gallons per thousand gallons of frack fluid; used to promote flow of the fluids used in the fracturing process.

Scale Inhibitor (ethylene glycol): seldom used; used to control the precipitation of specific carbonate and/or sulfate minerals.

Proppants (sand, resin coated sand, or man-made ceramic particles): usually approximately 1%-1.9% of total volume; used to hold fissures open so gas and oil can be extracted.

Now, many readers that have seen this type of information before are now expecting to see one of those "other uses" tables stating that fracking fluid must be safe due to the "ingredients" of fracking fluids having everyday uses, such as scale inhibitors having the same chemicals as windshield washer fluid, friction reducers having same chemicals as many makeup products, surfactants being basically the same as shampoo products, proppants being play sand, and hydrochloric acid also being swimming pool cleaner. These may be true in the strictest sense of the word, but this type of listing can also be very misleading and insincere, in that most all chemicals can be used for many different things, but are still not something with which one necessarily wants to come in contact. For example, ammonium nitrate is commonly used in agriculture as a high-nitrogen fertilizer, nitromethane is a commonly used industrial solvent, and Ryder trucks are commonly used to move families and their belongings to their dream homes; these are also three of the common "ingredients" used in the tragic April 1995 Oklahoma City bombing of the Alfred P. Murrah Federal building that killed 168 people. This is, of course, a comparison made for shock value, but it is meant as such to stick in your memory as how these sorts of comparisons can be manipulated and to drive home the fact that the best policy is to study upon facts when reading a comparison like this and make an informed decision for oneself.

9.3 Typical Percentages of Commonly Used Additives

Fracturing fluids are varied to meet the specific needs of each location; however, evaluating the widely reported percentage volumes of the fracturing fluid components reveals the relatively small volume of different chemicals that are present. Overall, the concentration of various chemicals in most fracturing fluids is a relatively consistent 0.5% to 2%, with water and proppants making up the remaining 98% to 99.5%. Keep in mind, however, that a typical fracturing job uses upwards of five million gallons of fracturing fluid, so a small percentage amount may actually result in a great deal of chemical usage, no matter how diluted it may be.

As one can imagine, the overall composition of fracturing fluids varies among companies and the drilling location. Fracturing fluids typically contain:

Approximately 90% water

Approximately 9.5% proppant materials

Approximately 0.5% chemicals: this percentage varies, but is typically between 0.5–1.0% by weight of total fluid

As described in previous sections, the chemical additives are included in fracking fluids to tailor the fluids to the requirements of the specific geological situation. The very popular chart below, taken from *Modern Shale Gas Development in the United States,*

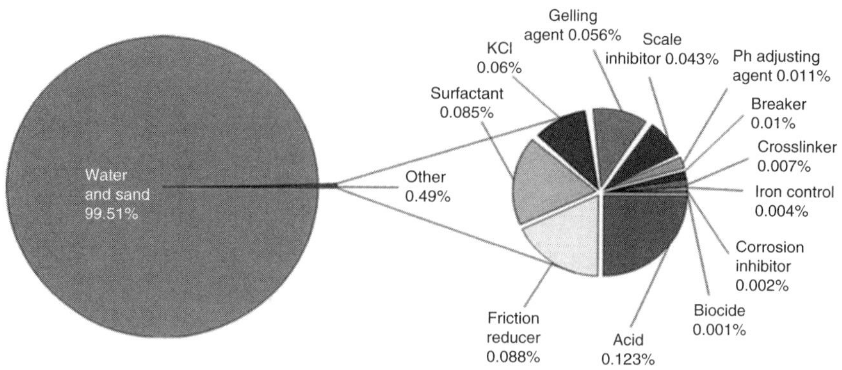

Graph 9.1 Volumetric Percentages of Additives in Fracturing Fluids.

demonstrates typical volumetric percentages of additives that were used for a typical hydraulic fracturing treatment of a shale horizontal well.

9.4 Proppants

Proppants are pretty hard to make into anything fun, exciting, or entertaining…as they are, for the most part, made up of sand or a manufactured facsimile of sand. Sure, if one wants to be poetic, one can refer to proppants as the only materials the operators want to remain downhole in the fractures. If one really wants to think poetically, feel free to consider a proppant's life as one of making its way from origins mined within the earth, only to return to its final resting place deeper within the earth's fractures.

As discussed earlier, proppants are simply materials (typically silica sand, resin coated silica sand, or manufactured ceramics) used to prop open the open fractures to promote flow and eventual extraction of hydrocarbons. As simple as proppants may seem, the estimated amount of proppant used in industry has grown tenfold since 2000. In some regions, it is not uncommon to see upwards of four million pounds of proppant used per well, and for proppant to represent up to 5% of well costs. The growth in proppant usage is generally attributed to operators realizing better well completion techniques with more proppant per stage and better well pad techniques with more laterals and fracturing stages per pad.

Even considering the accelerated growth in the last decade, the evolution of proppant usage has been slow to develop over the industry lifetime as a whole. Consider that the first frack job was conducted in 1947, utilizing a reported approximate 20,000 pounds of uncoated frack sand, and manufactured ceramic proppant was not first used until 1983, or 36 years later. Then, approximately one year later, resin coated proppant was first introduced. As with most all technologies, as new techniques continue to develop, proppants will surely evolve further to increase effectiveness and efficiency in hydraulic fracturing.

No matter the type of proppant used, the most important characteristics for a proppant are particle size distribution, crush resistance, shape, and sphericity (or roundness). Proppant materials are carefully sorted for size and sphericity to provide an efficient conduit for production of fluid from the reservoir to the wellbore. Grain size is critical because a proppant must reliably fall within certain size ranges to coordinate with downhole conditions and completion design.

Proppant shape and hardness qualities are also very important to the efficiency and effectiveness of a fracturing operation. A coarser proppant allows for higher flow capacity due to the larger pore spaces between grains, but it may break down or crush more readily under high closure stress, and rounder, smoother proppant shapes allow for better permeability.

Another important quality that must be taken into consideration is the proppant's hardness with respect to the formation. If the proppant is unable to embed in the formation, something referred to as point load occurs, which leads to higher flow capacity, but the

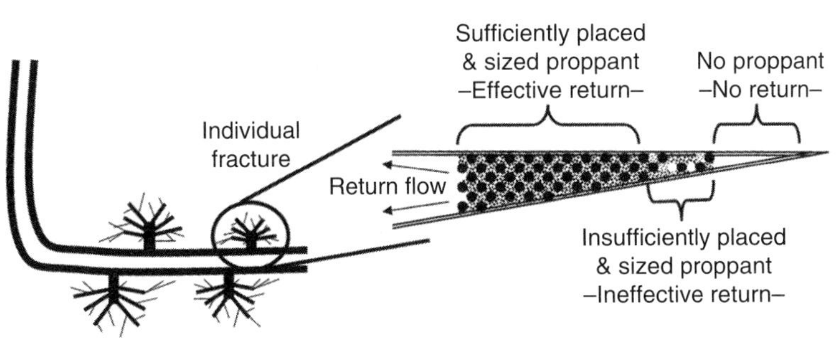

Figure 9.1 Proppant Size and Placement.

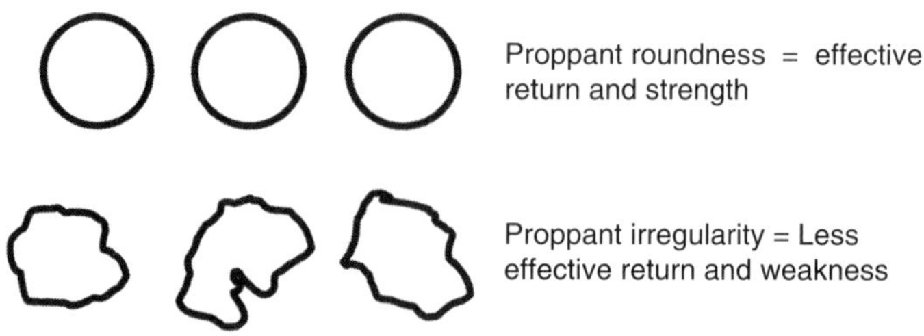

Figure 9.2 Proppant Shape.

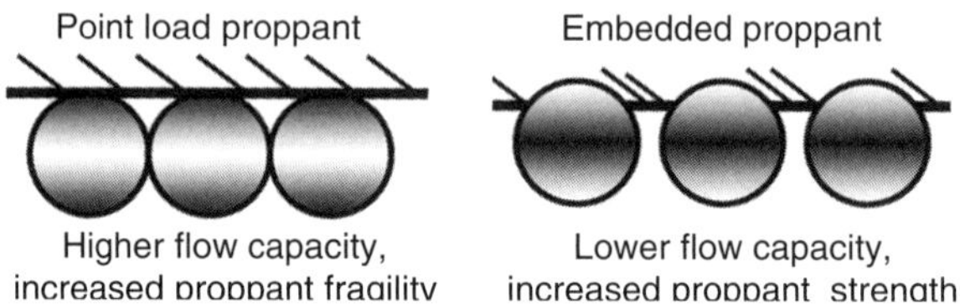

Figure 9.3 Proppant Hardness.

proppant will break more easily. However, if the proppant is able to embed in the formation, it is referred to as embedment, which results in the load pressure spreading out over the proppant area, increasing the breaking point, but also lowering flow capacity. Embedment is also a function of particle size.

Even though almost all proppant materials are naturally occurring, including manufactured ceramic proppants, with relatively low amounts of additional engineering necessary, the logistics in procuring and transporting proppants can be daunting. Logistical considerations include coordination of manufacturing material resources, transportation costs, and possibly a substantial monetary investment, or in equipment necessary, for processing and material handling facilities.

9.5 Silica Sand

While the all-encompassing term for the material "sand" is generally used for nearly all forms of broken down granules of minerals or rocks, to be specific, sand falls between silt and gravel in the spectrum of sizes. There are, however, many varieties of sand in the world, each with their own unique composition and qualities.

Everyone likes to picture the white sandy beaches of vacation destinations, for example, which are made up primarily of limestone that has been broken down. Then there are also many black sands, either volcanic in origin or containing magnetite. Other sands have high levels of iron in them, and so are rich and yellow in color.

The type of sand utilized for proppant materials is silica sand, which is, by far, also the most commonly used type of proppant. Silica sand, unlike many other "ingredients" of frack fluid, is more of a natural resource than an engineered product. Silica sand proppant is, in a simplistic description, made up of the most common mineral in the earth's continental crust, quartz. Silica sand is simply quartz that over the years, through the work of time and several erosion forces, has been broken down into tiny granules. Even though silica sand is a relatively common material, silica sand used for proppant is a specifically selected and utilized product. Proppant quality silica sand is a direct function of both the original depositional environmental and some slight mechanical processing, if necessary. Silica sand used for proppant is chosen for its round spherical shape and commonly graded particle distribution…unlike the common sand one might find at the beach or on a playground, which often feels gritty when rubbed between the fingers.

In addition to the oil and gas industry, there is some competition between other industries for the bulk of silica sand, as industrial-grade silica sand has a wide range of uses. This resource is also commonly used in the manufacture and preparations of various types of glass, in water filtration, sand blasting, as fill and as an ingredient in industrial concrete, in the metal casting industry to make cores and molds, and ironically, it is also used in the creation of highly flame-resistant industrial molds and construction materials for the kilns used in the manufacture of the sintered ceramic and bauxite proppants.

Even considering all the helpful and positive uses for silica sand across several different industries, there are some possible hazards related to its use. Because of the fine grains involved in silica sand, it can present a health risk if not properly handled. Care must be taken to keep the silica sand out of the lungs during use, and all materials containing more than 0.1% of silica sand must be clearly labeled. Workplace health applications also need to be in place and enforced: failure to wear a proper respirator or mask can result in lung irritation, and prolonged exposure can cause a chronic condition known as silicosis. Silicosis is a form of lung disease resulting from occupational exposure to silica dust over a period of years, causing a slowly progressive fibrosis of the lungs, impairment of lung function, and even a heightened susceptibility to tuberculosis of the lungs. Silicosis can also progress and worsen even after someone is no longer exposed to the silica dust, causing long-term effects and shortness of breath years later. Also, in the year 2000, the World Health Organization determined that crystalline silica is "associated with silicosis, lung cancer, and pulmonary tuberculosis" in classifying it as a Group I carcinogen "based on sufficient evidence of carcinogenicity in humans and experimental animals."

9.6 Resin Coated Proppant

As the name suggests, and to describe in the most simplistic of terms, resin coated proppant is exactly that: silica sand coated with resin. Resin coating silica sand proppant is utilized for two main functions:

1. to spread the pressure load more uniformly to improve the crush resistance of the silica sand particles, and
2. to keep pieces together that were broken from high closure stress from down hole pressure and temperature: this not only prevents broken pieces from flowing into the borehole, but also prevents these same broken pieces from returning to the surface during flowback production operation.

Currently, there are two types of resin coated proppants, *Pre-cured* and *Curable*. Pre-cured is the "original" technology, in which the resin coating on the silica sand grains is fully cured prior to injection into the fractures. The newer, curable technology has often been described as having a coating that is not completely "baked" or hardened. Curable resin coated proppants are used at a little more than half cure, so that when the proppant is pumped downhole it can finish curing in the fractures with down hole pressure and temperature. The advantage to curable proppant technology is that it allows the individual proppant grains to bond together in the fracture, resulting in the grains bonding together uniformly in strength when temperature and pressure reach appropriate levels.

9.7 Manufactured Ceramics Proppants

A third commonly used type of proppant is the manufactured ceramic proppant. This is a proppant generally manufactured from a type of ceramic material, typically non-metallurgic bauxite or kaolin clay. Bauxite is an aluminum ore from which most aluminum is extracted, while kaolin is one of the most common minerals, occurring in abundance from chemical weathering of rocks in hot, moist climatic soils like tropical rainforest areas.

Both bauxite and kaolin are utilized as proppants because of their superior strength characteristics, which are further enhanced through a process known as sintering. The sintering process is conducted in high-temperature kilns that are used to bake the bauxite or kaolin powder after it has been made into specifically sized particles. This process decreases the water content in the bauxite and kaolin to make them more uniformly shaped for size roundness and spherical shape. The desired results of this process are that the manufactured ceramic proppants can be engineered to withstand high levels of downhole pressure (closure stress).

9.8 Additional Types

As more is learned through the ongoing processes and further advances are made in technology, additional types of proppants are sure to come up. One current trend is toward the usage of "waste" material, including glass, metallurgical slags, and even rock cuttings produced to the surface during oil and gas drilling. The reuse of rock cuttings from gas drilling operations is especially attractive, since not only does it reuse a common waste product in industry, but it is also utilizing sources indigenous to the locality, which will cut down wastes while also cutting down on transportation and overhead costs. However, the other possibilities are also quite attractive, in that agreements can be made with landfills, metallurgical operations, and glass companies to recycle and reuse their wastes in lieu of land filling.

9.9 Slickwater

Slickwater is a more dilute, predominately water-based fracturing fluid that utilizes a limited amount of additive sand, friction reducers, and other chemicals. Slickwater is prepared this way in order to create a lower viscosity fracturing fluid to allow for an easier escape out of the created hydraulic fracture. Due to its low viscosity, slickwater can also be pumped down the wellbore at a higher rate.

10

So Where Do the Frack Fluids Go?

As discussed many times in this book, hydraulic fracturing involves the use of pressure applied by a fluid to create a new fracture or enlarge natural fractures present in the rock. These fractures extend in a planar style in general directions perpendicular to the plane of the least principal stress in the rock, following what is called "fracture direction." These hydraulic fractures are most commonly vertical (until contact is made with a rock of different structure, texture, or strength to stop the fracture height growth), and may extend laterally several hundred feet away from the well bore. These fracture barrier rocks, which stop the fracture upward or downward growth, are very common. Fracture height growth may extend up to a few hundred feet or more, but will likely be quickly limited by one of the dozens of rock barriers above the pay zone. Driving a fracture upwards through several thousand feet of rock is simply not possible, given the limits imposed by natural barriers and stresses of the formation and rocks above the pay zone. A good way to consider this is that if oil and gas are still in the reservoir millions of years after being created, then logic would dictate the reservoir barriers would also prevent the upward migration of fracturing fluids.

Pumping of a single frack stage may be limited to anywhere from 20 minutes to four hours, depending on the design and intent of the frack. This period of high pressure operation is normally the only time most wells will experience pressure high enough to reverse fluid flow into the formation. However, the required travel time for fluid to flow from the shale to aquifer under those pressures would be years. Any flow of frack fluid toward an aquifer through open fractures would be reversed during flowback.

As far as where the chemicals in frack fluids go, chemicals returning from a well after fracture are a fraction (usually between 20% or less for chemicals and about 40% for polymers) of what was pumped down the well. Most corrosion inhibitors, scale inhibitors, and surfactants adsorb onto minerals such as clays in the formation. Those that do not adsorb may bleed back very slowly, usually in a concentration of approximately 5-10 parts per million of water. The hydrochloric acid is typically used up within inches of the frack entry point and yields calcium chloride, water, and a small amount of CO_2; therefore, no live acid is returned to the surface.

11

Common Objections to Drilling Operations

The common objections to drilling operations is different than other objections for two reasons:

- There is no attempt to separate the operations related to fracking from all other drilling operations: this is simply because most nuisances related to one operation are the same for all (for instance, additional traffic is additional traffic, no matter the origination).
- The nuisances described in this section are not written of or discussed in a quantifiable way: in other words, this discussion is not centered on an amount, but the simple fact that it exists.

A reason for this is because a lot of data is collected for the other aspects related to fracking operations to attempt to prove/disprove their existence, when the nuisances discussed in this section are easily seen as being in existence (just spend a few minutes along any road used for drilling operations, and this will become abundantly clear). Please keep in mind that there are many additional nuisances absorbed by those living near drilling locations or related roadways, so this listing is far from comprehensive. The following are merely what most see as the most common, and are not presented in any order of magnitude. That decision has to be made by each individual: one person may be more affected by noise, while another is much more concerned with dust.

11.1 Noise

Noise conditions are usually one of the first things to change and be noticed by local landowners. An increase in noise is also one of the most continuous nuisances related to operations. Drilling and completing a well, from the pad construction to the final completion of the well, takes several weeks and utilizes many different types of equipment. This additional equipment can include additional trucking, construction, and drilling equipment. The noise concerns usually begin with the additional traffic brought to an area during pad construction, then continue with the noises associated with equipment and trucking required to construct a pad, only to be followed by the large amount of noise related to rig construction and operation throughout the well drilling process. Then, once the well site is completed, there may come the additional sounds of compressors used during ongoing production activities.

When one thinks of noise concerns related to the oil and gas industry, the first thing that commonly comes to mind is the big noisy rig, or maybe the noisy traffic coming back and forth. These are, of course, very real and valid concerns; however, the thing that is quite possibly the most notable noise nuisance related to the oil and gas industry, due to length of time, is the compressor. For the most part, the heavy rig work and heavy truck traffic lasts approximately one to two months, while the compressor, while not as loud, can continue for a much longer amount of time (months to even years).

Gas compressors are normally the largest equipment remaining after the well development process is complete and are utilized for something called gas lift. Gas lift is used in wells that have insufficient reservoir pressure to produce efficiently on their own. The gas lift process involves injecting gas through the tubing-casing annulus to aerate the fluid to reduce its density. Following aeration of the fluid, the formation pressure is then able to better lift the oil column up the wellbore. For pad sites where long-term compressor use is anticipated, especially in rural communities where serenity is the norm and even the slightest ongoing noise can be heard clearly for long distances, operators have addressed compressor noise concerns with remote siting (trying to locate the compressors on the part of the pad farthest from homes), noise tampering sound walls, and directing compressors with fans away from homes. However, even with the measures presently taken to mitigate ongoing sound issues, additional work must be done and technology developed to work toward a solution.

11.2 Changes in Landscape and Beauty of Surroundings

Several different types of pollution are commonly mentioned in relation to the oil and gas industry, including water pollution, spoil pollution, air pollution, and, as presented in the previous section, noise pollution. However, one that may be overlooked to the majority of the public, but certainly not overlooked to those affected, is visual pollution. Visual pollution is an aesthetic issue, referring to the impacts of pollution that impair one's ability to enjoy a vista or view. Now, with the possible exception of the immense number of billboards lining the nation's highways, not many things meet the definition of visual pollution as much as a drill rig.

Drill rigs utilized in most unconventional well drilling typically can range from approximately 50 feet to 100 feet in height. Couple the height of the drill rig with the ongoing movement and dust related to drilling, and it is easy to imagine how this would be bothersome to those adjacent to rig locations. One mitigation attempt for this problem would be the usage of lower height rigs. However, the undesirable trade-off for a lower height rig is the necessary extended time on location for smaller rigs.

Ironically, horizontal drilling techniques commonly related to unconventional well drilling and hydraulic fracturing locations can actually be considered a "semi-solution" to this problem. Pads used for horizontal drilling commonly include multiple laterals on one location, in which the drilling of multiple wells literally means moving the rig over as little as twenty feet from the original location. This allows wells to be drilled from one location without the necessity of moving the rig and drilling in several locations, which would only disturb that many more possible visual pollution points. This also allows for accelerated drilling time due to lessened rig movement time, a reduction in the number of necessary lease roads and drill pad locations, fewer necessary pipelines, and fewer tank batteries.

11.3 Increased Traffic

Another nuisance commonly cited by those living in oil production areas is the drastic amount of added traffic it creates. This is not necessarily the type of traffic most think of when hearing the word. Traffic related to the oilfield includes all of the initial traffic to bring in heavy equipment for pad construction and eventually the rig itself, followed by traffic for well completion and fracking activities (to get a taste for what this is like, consider the amount of sand used in each frack job, then consider how many separate truckloads that would be), then the ongoing traffic related to hauling produced water and oil from the locations until some sort of pipeline infrastructure can be put in place.

Also, keep in mind that many of the areas affected by oil and gas operations are rural and do not, quite simply put, have the proper roadways for the larger size or amount of traffic vehicles that come with industry operations. Not only does increased traffic add additional wear and tear to the local roadways, the narrower two-lane and sometimes even more narrow gravel roads cause very unsafe driving conditions for the industry and local resident vehicles alike.

The answers to the traffic problems may seem obvious – do something to lessen the amount of traffic or do something to improve the roads – but finding ways to turn those answers into reality is something much more difficult than may first appear. The first possible answer, "do something to lessen the amount of traffic," would include:

1. the need to either use fewer (but larger) transport vehicles, resulting in additional hazardous conditions with the larger vehicles on the narrow rural roads; or
2. the need to install a pipeline infrastructure to transport produced water and/or oil, which comes with the obvious concerns related to pipelines installation and location.

The second possible answer, "do something to improve the roads," would depend on the type of road to be improved. Improving and widening gravel-type lease and rural roads is a less daunting task than improving paved city/county roads due to ease of obtaining the proper materials and fewer restrictions put on maintenance. However, making improvement to city and county roads would include needing to clear a wider right-of-way all along the road to be widened, and the city/county would need to have the funds set aside for this task, which is a time consuming process.

11.4 Subsurface Contamination of Ground Water

An increasingly common concern expressed about hydraulic fracturing is that operations create fractures extending past the target formation to aquifers, allowing fracturing fluids to migrate into drinking water supplies. There are several factors that would prevent this from occurring:

- This would require hydro-fractures to extend several thousand feet beyond the upward boundary of the target formation through many layers of rock.
- Geologists estimate there is at least a half mile of rock between the natural gas deposits and the groundwater.
- Impermeable shale acts as a barrier to vertical propagation of both natural and artificial fractures.
- After hydraulic fracturing is completed, the fluid flow is toward – not away from – the well as gas enters the well bore during production.

Despite what the protests may lead one to believe, as discussed before, studies have shown that failure of the cement or casing surrounding the wellbore poses a far greater risk to water supplies than hydraulic fracturing.

11.5 Impacts on Water Wells

Many allegations have been made by residents and documentaries of impacts on water wells that hydraulic fracturing activities are causing some of the most visible and contentious controversies. Most allegations of water well impacts involve methane, color changes, turbidity, and odor.

Particularly in areas underlain by gas-producing shales, methane migrates out of the shales under natural conditions and moves upward through overlying formations, including aquifers. Such naturally-occurring methane in water wells has been recognized as a problem in shale gas areas for many years or decades before shale gas drilling began.

11.6 Water Analysis

The quality of groundwater can affect not only health, but also society and the economy. Groundwater contamination can adversely affect property values, the image of a community, economic development, and the overall quality of life shared by all. Clean water at a reasonable cost is essential, and in many parts of the country, groundwater is the only economical water source available. Once groundwater has been contaminated, it is usually very difficult and costly to clean. Even small contamination sites often cost many thousands of dollars to clean up. The quality of water from private water supplies, such as those from wells at individual homes, is not regulated. It is the responsibility of the well owner to ensure a safe drinking water supply. Although there are a few requirements for water quality testing and monitoring of private wells (i.e., in some areas, testing is required at the time of property transfer), it is recommended that all well owners have their water tested periodically. While "complete" drinking water analyses can be expensive and are generally unnecessary for the private well owner, it is recommended that private water supplies be tested routinely for common contaminants including total coliform bacteria, nitrates, and lead. These contaminants can occur in well water due to agricultural activity, septic system use, household chemical use/ disposal, age of the plumbing, or industrial activity. The frequency of water testing and the contaminants to test for depend on factors such as the potential sources of pollution and the type of well. Another consideration is ensuring that the private well complies with proper well construction standards.

A water's taste, smell, or color is not necessarily an indicator of water quality. Many of the most hazardous contaminants are undetectable to the senses. The only way to detect most pollutants is by testing.

Before hydraulic fracturing operations begin in a new area, American Petroleum Institute guidance (API - HF1) recommends that a baseline assessment program that includes the sampling of nearby water wells be conducted prior to hydraulic fracturing operations. Fresh water wells should also be sampled following hydraulic fracturing operations. At least one state (Colorado) requires the sampling of certain water wells in various areas of the state as part of their regulatory program. Another state (Pennsylvania) has regulations that presumptively place the burden of proof on any oil and gas company to demonstrate that they have not caused deterioration of the quality of groundwater used for drinking water purposes in the vicinity of oil and gas wells in the event of a contamination complaint.

In order to obtain valid results from sampling, it is important to follow proper sampling and analysis protocols. Contact a state or EPA-certified laboratory for sampling containers and instructions. Proper protocols may include:

1. using appropriate containers and seals;
2. purging of the well prior to sample capture;
3. collection at points before water treatment equipment;
4. following sample container filling procedures;
5. following storage and holding time requirements;
6. utilizing appropriate analysis methods; and
7. following appropriate quality control/ quality assurance protocols.

Sampling should be conducted by someone familiar with sampling procedures. Analyses should be conducted by an accredited laboratory using appropriate analysis methods. One may be able to obtain a list of qualified laboratories by contacting a local health department, state water quality agency shown on the regulations by state page, or county extension agent.

It is important for the landowner to have an oil and gas operational sampling and analysis of their groundwater conducted by a professional for constituents that may provide a reasonable baseline for post-fracturing analysis. The National Ground Water Association maintains a list of groundwater professionals that the reader can review to help find someone in the local area for assistance. The following is a good basic list of constituents that should be considered for analysis prior to oil and gas operations.

- Major cations and anions
- pH
- Specific Conductance
- Total Dissolved Solids

- Benzene, Toluene, Ethyl benzene, Xylene (BTEX)/ Diesel Range Organics (DRO)/ Gasoline
- Range Organics (GRO)
- Total Petroleum Hydrocarbons or Oil & Grease (HEM)
- Arsenic
- Barium
- Calcium
- Chromium
- Iron
- Magnesium
- Selenium
- Boron
- Sodium
- Chloride
- Potassium
- Bicarbonate
- Dissolved Methane

Once hydraulic fracturing has taken place and a record of the actual chemicals used is available, it would be advisable to consider having a sampling and analysis conducted on the groundwater for the chemicals shown on the record that match those listed above or those that are by-products, reaction products, or daughter products of those listed above. This is important because many of the chemicals used in hydraulic fracturing will be degraded, oxidized, or otherwise modified during the fracturing process. Thus, simply looking for the chemicals on the list shown above may not yield enough information for a comparative analysis. However, to minimize costs for the landowner, an alternative analysis should be conducted for at least Total Dissolved Solids (often abbreviated TDS) is a measure of the combined content of all inorganic and organic substances contained in liquid and dissolved methane. An increase in the concentration of either of these constituents could indicate that further, more complete sampling and analysis should be conducted.

The reader can learn more about the toxicity characteristics of chemicals by searching for the chemical using the name or CAS number on the USEPA National Center for Computational Toxicology website. USEPA also maintains a Drinking Water Hotline that is available Monday-Friday from 8:30 AM-4:30 PM Eastern Time at 1-800-426-4791.

Important: Only a trained professional such as a toxicologist or a physician can tell you if your water is safe to consume. The reader should not use the information obtained from the USEPA Toxicology website, or any other website to make decisions regarding the safety or drinkability of water.

11.7 Types of Methane and What They Show Us

One of the most graphic and widely publicized issues related to hydraulic fracturing is the thought that it causes methane gas to emit from water wells, in some instances in ignitable concentrations. However, methane has always been a relatively common contaminant in water wells and can be the result of natural or man-made causes. Methane is a natural hydrocarbon gas that is flammable and explosive in certain concentrations and is produced either naturally by bacteria or by geologic processes involving heat and pressure.

There are several possible causes for methane gas to be found in water wells unrelated to hydraulic fracturing operations. These include naturally forming biogenic methane caused by the decay of organic materials or natural seeps of thermogenic methane that have been coming to the surface for millions of years.

Biogenic and thermogenic methane are reasonable easy to differentiate through analyses, and through this "fingerprinting," experts can determine if specific instances of methane in a well are related to biogenic or thermogenic processes. Biogenic and thermogenic methane differ in the carbon isotopes they contain, with biogenic methane containing more C^{12} carbon, while thermogenic methane contains more of the C^{13} carbon isotope. Biogenic methane is also nearly 100% methane, while thermogenic methane may also contain some propane and butane from thermal decomposition.

11.8 Biogenic

Biogenic methane is a near-surface, naturally occurring microorganism type of methane that is continually produced as a by-product from decay of organic materials by reactions in wetlands, sewage, landfills, and even agriculture. Once formed, it will often remain in some shallow, water-bearing geologic formations, into which many domestic water wells are completed.

11.9 Thermogenic

Unlike biogenic methane, thermogenic methane is formed deep within the earth from the thermal decomposition of buried organic materials deposited millions of years ago. It is found buried deeper within the earth, and thermogenic gas found near or at the surface is a strong indicator of natural seeps.

11.10 Possible Causes of Methane in Water Wells

As mentioned earlier, many water wells are liable to produce some level of odorless methane through sources such as agriculture, livestock, wetlands, landfills, and other sources allowing biogenic gas to form through decaying or vegetation. In addition to these natural causes, it is also quite possible to have detectable methane content in a water well caused by a nearby improperly constructed gas or oil well. A third possible cause of methane in water wells is from the penetration of coal formations that are saturated with fresh water. Coals can have as much as 90% or higher organic content and gas that naturally adsorbed on the organic materials in the coal can desorb as water is pumped from the area, such as water well.

Proof of naturally occurring methane near surface in concentrations high enough to allow ignitability can be seen in natural seeps. There are literally thousands of natural oil, gas, and salt water seeps that come to the surface. The earliest historical written record of natural oil and gas seeps in the Western Hemisphere was that of Sir Walter Raleigh's account of Pitch Lake in Trinidad in 1595. Well-known current oil and gas seeps in the United States include the La Brea Tar Pits and McKittrick Tar Pits in California and the "Eternal Flame Falls" of the Shale Creek Preserve, a section of the Chestnut Ridge Park in New York.

11.11 Surface Water and Soil Impacts

Another major concern for hydraulic fracturing well production is the potential for spills or releases at the well pad site or during transportation. Potential sources of release are fuels and oils for the drilling rig and on-site equipment, possible overflow or leak of storage tanks and pits, releases or spills of drilling mud, and flowback, produced water, or hydraulic fracturing fluid released or spilled during storage. In contrast to the effects they may cause downhole, chemical additives pose quite a higher risk in their concentrated form while being transported or stored on site than when injected into the subsurface during hydraulic fracturing.

Common sources of spills at the pad site include the drill rig itself from possible mechanical failure and other operating equipment, storage tanks, pits, and even leaks or blowouts at the wellhead. Leaks or spills may also occur during transportation (by truck or pipeline) of materials and wastes to and from the well pad. As in most all of these cases, surfaces water and groundwater are the primary risk receptors. Root causes of on-site and off-site releases can be accidents, inadequate facilities management or staff training, or even illicit/illegal dumping.

In the event of spills, effective containment is a major factor in minimizing the impacts on human health and the environment when a spill occurs. Containments, including berms around tank batteries, double walls in tanks and vessels, and catch pans and basins for smaller storage containers, are the first line of defense against migration into exposure pathways that are linked through surface water or groundwater to humans, animals, or other receptors.

11.12 Spill Preparation and Documentation

As will be discussed later in this section, the Clean Water Act addresses spills and other accidental releases. A main component of spill prevention is the Spill Prevention Control and Countermeasures (SPCC) plan prepared to assure that adequate responses, materials, and personnel are available to respond to releases. Most states require an SPCC plan or equivalent for oil and gas operations as well as a statewide plan for response to spills and other releases. SPCC plans include prevention and control measures along with contact information, site-specific maps and details, reporting and cleanup requirements, and specifications for proper materials handling.

11.13 Other Surface Impacts

Drilling operations require significant above-ground development. In addition to possible spill scenarios related to the well pad itself, other possible surface impact scenarios are also quite common.

Lease roads may need to be built and gathering infrastructure installed to bring the product from the wellhead to a delivery point or a pipeline. The pipeline itself, including installation and the remaining easement, may require the development of several acres of land.

Total land use and resulting impacts to the environment can be reduced by drilling multiple wells from a single well pad, which ironically is a very common practice on horizontal hydraulic fracturing well pads. Nonetheless, with so many wells being drilled and the related infrastructure development, it is imperative that industry and regulatory bodies do as much as possible to minimize the overall impact on local communities, habitats, and road systems.

11.14 Land Use Permitting

Land use decisions affect a wide range of stakeholders, including landowners, adjacent neighbors, surrounding communities, and community leaders. Permitting procedures will need to consider the needs of each of these stakeholders and include concise and enforceable preventative and remediation strategies to help provide minimal impact and maximum restoration of the land associated with well pads and production. Proper oversight also needs to be provided to help manage soil erosion and transport of sediment into streams and other water bodies, prevent damage to ecological habitats, and avoid fragmentation of habitats.

During the construction phase for a well pad or infrastructures, the quality of surface water resources may be impacted by runoff, particularly during storm events. Shale gas development will also affect forests and ecological habitat on a large scale. Development studies indicate that many well pads are constructed in forest clearings, resulting in the clearing of thousands of acres of habitat from pads and associated road infrastructure.

11.15 Water Usage and Management

Flowback water is the portion of injected fluid that returns to the wellbore after the downhole hydraulic fracturing process is completed, and the fluid pressure is relieved. Flowback water is

Table 11.1 Water use comparison.

Energy Resource	Range of Gallons Water Used per MMBTU of Energy Produced
Shale Natural Gas	0.60 – 1.80
Natural Gas	1 to 3
Coal (no slurry transport)	2 to 8
Coal (with slurry transport)	13–32
Nuclear (processed uranium ready to use in plant)	8 to 14
Conventional Oil	8 to 20
Synfuel-Coal Gasification	11 to 26
Oil Shale Petroleum	22 to 56
Tar Sands Petroleum	27 to 68
Synfuel-Fischer Tropsch (Coal)	41 to 60
Enhanced Oil Recovery (EOR)	21 to 2,500
Fuel Ethanol (from irrigated corn)	2,510 to 29,100
Biodiesel (from irrigated soy)	14,000 to 75,000

then brought back to the surface for treatment, recycling, and/or disposal. The initial fluid withdrawn from the wellbore actually consists of a mixture of the flowback water and saline water from the formation wellbore, referred to as "produced" water.

There is much concern over the use of water for fracking. Consider the following table:

The source for the above table is the Chesapeake Energy 2009 presentation to the Ground Water Protection Council, citing Chesapeake well estimates for shale gas and a U.S. Department of Energy water use report: Chesapeake gave the authors its sources on these facts. Coal plant water consumption: "Energy Demands on Water Resources," U.S. Department of Energy, December 2006, http://www.sandia.gov/energy-water/docs/121-RptToCongress-EWwEIAcomments-FINAL.pdf. NYC water consumption: New York City Department of Environmental Protection. Golf course consumption: Colorado State University Agricultural and Resource Policy Report, April 2004.

Chesapeake says that the water it uses to frack an average shale well is the same amount consumed by a coal-fired power plant in 12 hours. It is what New York City consumes in seven minutes. Even recreation compares unfavorably: a golf course drinks the same amount in 25 days and then drinks that same amount every month, year after year.

11.16 Flowback Water

Flowback water contains some or all of the following: sand and silt particles (from the shale or returned proppants), clay particles that remain in suspension, oil and grease from drilling operations, organic compounds from the hydraulic fracturing fluids and the producing shale, and total dissolved solids from the shale. The amount of injected fluid returned as flowback ranges widely – from 20% to 80% – due to factors that are not well understood. The flowback period can range anywhere from hours to weeks: as continued withdrawal proceeds, the flowback water becomes more saline as the relative amount of produced water increases.

Proper management of the flowback and produced water streams has been a major issue of the hydraulic fracturing controversy. Issues arise from the spills, both on location as well as during transport of these water sources to the proper disposal. Disposal of flowback water has historically been by permitted injection wells in some areas and by discharge to publicly-owned treatment works in others. However, discharge to publicly-owned treatment works, which is necessitated by less desirable subsurface conditions for underground injection wells in eastern states, has become controversial and has been prohibited by some states, while other states require pretreatment before discharge to a publicly-owned treatment works.

11.17 Produced Water

Produced water is generally a term used in the oil industry for water that is produced and returned to the surface along with the oil and gas. Produced water originates as a natural water layer that lies under the hydrocarbons in oil and gas reservoirs. Produced water is typically found in greater amounts in oil reserves than gas reserves.

11.18 Flowback and Produced Water Management

One of the major problems concerned with handling flowback water and produced water is the relatively large amount temporary storage followed by transport of fluids prior to treatment or disposal. Flowback water and produced water is commonly stored on a temporary basis in on-site pits or tanks before transport removal by truck or pipeline for reuse, treatment, or disposal. These pits or tanks are another possible source of leaks or spills. Lining of pits, or lack thereof, for flowback water and produced water depends on company policies and regulatory requirements, which vary from state to state.

Even in cases where liners are used, they may leak either from age and wear from use or even by improper installation and upkeep. These leaks result in releases to the subsurface, and have led many to question the use of pits in favor of closed-loop steel tanks and piping systems. Storing flowback water and produced water in enclosed steel tanks, a process many companies have already adopted, has led to a reduction in the risk of releases while improving water retention for subsequent reuse. The following chart is a breakdown of the various substances found in the produced water.

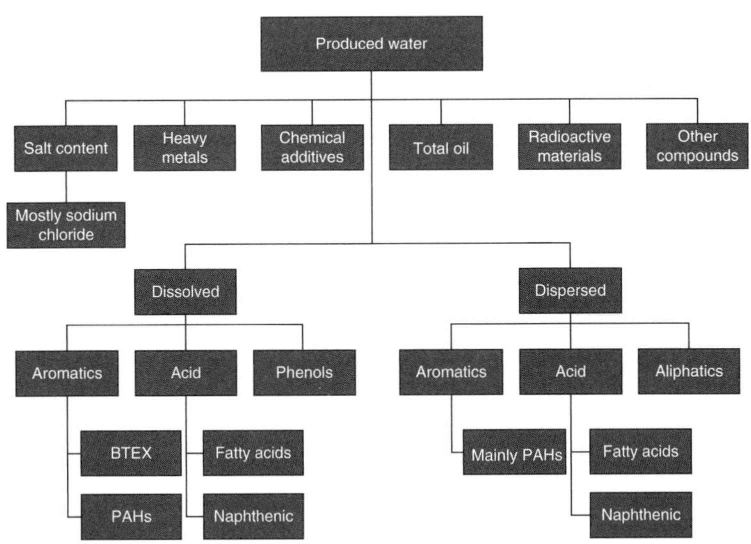

Chart 11.1 Water and Contaminates.

11.19 Geological Shifts

Contrary to popular belief, seismic events have been extremely common throughout our nation's history, and with improving technology and detection methods, the United States Geological Survey has compiled data showing that the number of damaging earthquakes has remained constant. Also, their data indicate that there appears to be no direct connection between hydraulic fracturing and damaging earthquakes. However, fracturing into a moderate size fault may produce seismic energy sufficient to create measurable signals at instruments very close to the frack.

Data has shown that the hydraulic fracturing process, while not necessarily creating damaging earthquakes, does create a large number of micro-seismic events, or micro-earthquakes. The magnitudes of these micro-seismic events are generally too small to be detected at the surface. The biggest micro-earthquakes have a magnitude of about 1.6: an event of this size represents the slip of about the thickness of a human hair or about as much energy as is released by a gallon of milk dropped from chest height to the floor.

11.20 Induced Seismic Event

Scientists have known for some time that pumping fluids in or out of the earth's subsurface has the potential to cause seismic events. These seismic events, when attributable to human activities, are called "induced seismic events." Data collected from research at several energy development sites has illustrated factors that induce seismicity, including:

- The presence, orientation, and physical properties of nearby faults
- The volumes, rates, pressures, and temperatures of fluids being injected or withdrawn
- The subsurface properties in the location

11.21 Wastewater Disposal Wells

Tens of thousands of injection wells have been drilled and operate in production areas to dispose of the flowback water and produced water generated by oil and gas production operations. This flowback and produced water injection for disposal has also been suspected and, in some cases, determined a likely cause for induced seismicity in the past several decades. Unlike production wells, those used only for flowback and produced water disposal normally do not undergo detailed geologic review prior to construction.

11.22 Site Remediation

The requirements for the removal, known as plugging and abandoning, of production wells at the end of their life cycle is determined by state agencies. States also have oversight for specifying site restoration requirements.

The common objective in site restoration for drill pads and other infrastructure requirements is to restore the site to its former conditions and use. Prior to release of the location for other uses, operators are required to test for contamination, clean up all apparatus and restore the location to prior drainage patterns, and reinvigorate the site within a reasonable time.

11.23 Regulatory Oversight

The different processes of shale gas development are regulated at almost all levels of government. However, the principal regulatory authority generally lies within each state. Not only is primary regulatory authority at the state level, many federal requirements have also been delegated to the state level. State agencies typically administer the federal environmental regulations and also are tasked with writing and enforcing their own regulations, governing nearly all phases of oil and gas operations.

In addition to problems encountered through differing states managing production facilities by differing regulations, most oil and gas regulations were written before hydraulic fracturing became the industry that it is today; therefore, it is subject to oil and gas regulations enacted prior to its development and in many cases not adequately addressing fracturing processes. In general, too few regulations are currently directed specifically to hydraulic fracturing: many applied regulations are directed to broader environmental laws for air, water, waste, and land development. However, state and federal regulations are currently being developed and and/or revised to focus on:

- Proper cementing and casing of wells
- Disclosure of hydraulic fracturing chemicals
- Proper management and disposal of flowback water and produced water

In addition to having concise and relevant regulations, it is also very important to have adequate enforcement of regulations both by office and field staff conducting field inspections. This includes having an adequate number of staff assigned, conducting inspections, and ensuring violations are recorded. The type and severity of violations demonstrate the type of adverse effects being addressed by the regulatory programs.

To be truly effective, regulations should focus on the most urgent issues, such as spill prevention, which may pose a greater risk of interacting with receptors than hydraulic fracturing itself. Most recorded violations are associated with overall gas drilling operations rather than being specific to hydraulic fracturing processes: surface spills, improper disposal of oil and gas wastes, problems with leaking pits or tanks, and administrative issues are typically the most common violations.

11.24 Federal Level Oversight

Hydraulic fracturing is subject to many federal regulations (as is the case for other oil and gas operations), but also receives a great number of exemptions that normally would have been applicable. This is partly to do with "process knowledge" associated with the industry, which means if processes are continuously conducted by the same protocols, then the same results with wastes and regulations; therefore, it can self-regulate. This is meant as a time- and resource-saving area for regulatory agencies as much as anything else. This has also led to cooperative efforts between regulatory agencies and the industry to optimize the effectiveness of regulations. However, in some states, similar requirements that are exempted from federal regulation are imposed at the state level.

11.25 State Level Oversight

As stated earlier, states have been delegated a great deal of oversight and management for hydraulic fracturing regulations and enforcement. However, with a great deal of regulations having been enacted prior to hydraulic fracturing becoming the topic it is today (particularly in states not having previous extensive oil and gas development), new or additional industry-specific regulations, such as chemical management requirements, may be needed to minimize surface and subsurface disturbances and impacts on environmentally sensitive areas.

To help alleviate this, a number of organizations and activities are underway, including the Groundwater Protection Council (GWPC) and State Review of Oil and Natural Gas Environmental Regulations (STRONGER), to research and develop state regulation of oil and gas operations.

11.26 Municipal Level Oversight

Municipal oversight is provided for mostly secondary processes related to oil and gas developments that are not regulated by state and/or federal entities: these include traffic management, road maintenance and repair, and usage and bridge zoning.

11.27 Examples of Legislation and Regulations

The following laws and regulations are associated with varying phases and protocols of shale gas development:

- Clean Water Act (CWA): Oil and gas operators must obtain a storm water permit under the Clean Water Act for the construction and operation of a well pad and access road that is one acre or greater. Also, the Clean Water Act prohibits the dumping of any pollutant into U.S. waters without a permit. Typically, facilities that may generate storm water runoff must obtain a storm water permit for this runoff. However, the Clean Water Act does not require oil and gas operators to obtain a permit for uncontaminated "discharges of storm water runoff from . . . oil and gas exploration, production, processing, or treatment operations."

- Clean Air Act (CAA): Recently-proposed Clean Air Act regulations are intended for operators to control volatile organic compound emissions from flowback during the fracturing process by using volatile organic compound capture techniques called "green completion."
- Endangered Species Act (ESA): Operators must consult with the Fish and Wildlife Service and potentially obtain an incidental "take" permit if their operations may affect endangered or threatened species by well development.
- Migratory Bird Treaty Act (MBTA): Operators are held strictly liable for any harm to migratory birds, and must ensure that maintenance of surface pits or use of rigs does not attract and harm these birds.
- Emergency Planning and Community Right-to-Know Act (EPCRA) and Occupational Safety and Health Act (OSHA): Operators must meet safety requirements in a myriad of work processes such as working at heights, tank entry, excavation, medical surveillance, first aid, and chemical storage. Operators must also maintain material safety data sheets for certain hazardous chemicals that are stored on site in threshold quantities.
- Comprehensive Environmental Responsibility, Compensation, and Liability Act (CERCLA): Operators must report releases of hazardous chemicals of threshold quantities and may potentially be liable for cleaning up spills.
- Resource Conservation and Recovery Act (RCRA): Most wastes from hydraulic fracturing and drilling are exempt from the hazardous waste disposal restrictions, meaning that states – not the federal government – have responsibility for disposal procedures for the waste.
- Safe Drinking Water Act (SDWA): Hydraulic fracturing operators also are exempt from the Safe Drinking Water Act, which requires that entities that

inject substances underground prevent underground water pollution. The SDWA applies only to waste from fracturing and drilling that is disposed of in underground injection control wells. If operators use diesel fuel in fracturing, however, they are not exempt from SDWA.

11.28 Frack Fluid Makeup Reporting

Of all the issues related to hydraulic fracturing, the issue of frack fluid makeup reporting may be the one that generates the most attention from state legislators and the concerned public. As discussed in earlier in this section, an overriding reason for this level of concern is greatly related to semantics and the inexplicable refusal of many companies to openly disclose frack fluid makeup sooner and with more sincerity.

The detailed composition of the additives has been controversial because, until recently, companies that manufacture fracturing fluid components have insisted the exact composition was proprietary, and some components should therefore not be reported. Over the last few years, however, a greater number of voluntary disclosures and state-based disclosure laws have resulted in increased disclosure in the details of frack fluid composition. In spite of a great deal of positive ground made in the pursuit of full disclosure of frack fluid ingredients, there is not yet a clear understanding of what the full list of chemicals or their impact on the environment might be.

In a helpful compromise between industry and the concerned public, recently enacted frack fluid make-up reporting regulations have stipulations that not only address the disclosure of chemicals used, but also provide some level of protection for proprietary trade secrets. The energy industry as a whole has supported this approach (see e.g., www.Fracfocus.org) and generally agrees the disclosure of the chemicals used in frack fluid makeup is appropriate and necessary; however, industry maintains that proprietary information should not be forced into publication.

Continued movement toward detailed disclosure of chemicals present in hydraulic fracturing fluid will enable a continued analysis of the chemicals' potential impact and will help address and alleviate public concern over their risk to water resources.

11.29 FracFocus

As stated on the website, FracFocus is a national hydraulic fracturing chemical registry managed by the Ground Water Protection Council and Interstate Oil and Gas Compact Commission, two organizations whose missions both revolve around conservation and environmental protection. The site was created to provide the public access to reported chemicals used for hydraulic fracturing within their area. To help users put this information into perspective, the site also provides objective information on hydraulic fracturing, the chemicals used, the purposes they serve, and the means by which groundwater is protected.

The stated primary purpose of FracFocus is to provide factual information concerning hydraulic fracturing and groundwater protection. It is not intended to argue either for or against the use of hydraulic fracturing as a technology. It is also not intended to provide a scientific analysis of risk associated with hydraulic fracturing. While FracFocus is not intended to replace or supplant any state governmental information systems, it is being used by a number of states as a means of official state chemical disclosure. Currently, eight states, Colorado, Oklahoma, Louisiana, Texas, North Dakota, Montana, Mississippi, and Pennsylvania, use FracFocus in this manner.

11.30 Atmospheric Emissions

Air emissions from shale gas operations occur at the drill site during drilling and fracturing and at ancillary off-site facilities such as pipelines and natural gas compressors. The on-site emissions include dust, diesel fumes, fine particulate matter (PM 2.5), and methane. Air emissions have become a major component of the shale gas controversies.

A principal concern for shale gas emissions is related to the volatile organic carbon (VOC) compounds. Depending on the composition of the gas produced from the shale, VOCs are typically rich in the BTEX (benzene, toluene, ethylene, xylene) compounds. However, the role of VOCs as smog precursors – they combine with NOx in the presence of sunlight to form smog – is the main source of concern with these compounds. Ozone, a primary constituent of smog, and NOx are two of the five "criteria pollutants" of the Clean Air Act (CAA). The Fort Worth area in the Barnett shale play has been designated "non-attainment" for ozone under the CAA, which means that the established standard is not met for ozone concentration in the atmosphere. The role of VOCs in forming smog and their contribution to the elevated levels of ozone is the reason for the focus on VOC emissions from shale gas activities.

However, the contribution of shale gas activities to ozone levels is highly controversial. For example, investigations in the Fort Worth area have found that most VOCs are not associated with natural gas production or transport. Allegations that VOC and NOx emissions from natural gas production from Barnett shale activities play a significant role in ozone formation have been strongly contested. Records of the Texas Commission on Environmental Quality (TCEQ) monitoring program since 2000 actually show overall decreases in the annual average concentration of benzene, one of the VOCs, during the period of early shale gas development in the Fort Worth area.

Public concern over air quality and the need for more precise information led to more focused emissions studies sponsored by local governments or private foundations. The first – and most controversial – of these studies was at DISH, Texas, where elevated levels of benzene, xylene, and naphthalene were found from a set of 24 samples and four residences. DISH was originally named Clark. In November 2005, the community accepted an offer to rename itself "DISH" (all capital letters) as part of a commercial agreement with a satellite television company. Another study in a very active area of shale gas production located about seven or eight miles from DISH found that shale gas was responsible for less than half of the VOCs (43%) in the atmosphere, with motor vehicle emissions contributing most of the rest (45%).

Modeling studies indicate that 70 to 80% of benzene is from fugitive emissions of natural gas, but that other VOC constituents are from motor vehicle emissions. In portions of Western states such as Wyoming, air emissions from oil and gas activities are the largest source of VOCs and related high ozone levels. In Sublette County, Wyoming, for example, ozone levels in the winter routinely exceed the EPA 8-hour standard, resulting in air quality that is sometimes worse than in Los Angeles.

Allegations that the emission of VOC constituents such as benzene in "widespread" or "prevalent" amounts in shale gas operations appear not to be supported when comparisons are made with air quality standards or when the relative amounts are compared to other sources such as vehicle exhausts. The relative contribution of shale gas activities in relation to conventional oil and gas development and other sources such as vehicle exhaust emissions must be taken into account in reports such as those from Wyoming and Fort Worth.

Emissions of methane have caused public concerns over global climate change, since methane is a strong greenhouse gas. Venting or flaring of natural gas may take place during the fracturing and flowback phase of shale gas well development. However, many operators use "green completions" to capture and sell rather than vent or flare methane produced with flowback water. Onsite fugitive emissions of methane may take place from other sources as well, such as pressure relief valves of separators, condensate tanks, and produced water tanks. Although natural gas is confined in pipelines from production wells to the point of sale, methane emissions may also occur from offsite gas processing equipment and compressors notwithstanding the economic motive to minimize loss of natural gas. It is not known in the public realm the extent to which Best Management Practices, which is an industry term, (e.g., low-emissions completions, low-bleed valves) result in reduced methane and fugitive losses of methane.

12

Air Emissions Controls

Shale gas development is subject to both federal and state air emissions regulations established by the Clean Air Act (CAA) and state-level legislation. Many of the CAA provisions are delegated from the U.S. EPA to the various states' environmental agencies. The major air pollutant sources of shale gas drilling and fracturing are the drilling and associated equipment, tanks and pits for flowback water, flared gas, and methane sources at the wellhead and from pipelines and compressors.

Oil, gas, and shale gas operations in particular are subject to regulations for "criteria pollutants" (sulfur dioxide, nitrogen oxides, ozone, particulate matter, carbon monoxide, and lead) and "hazardous air pollutants" (HAPs, including 187 compounds). However, these regulations focus on "major" sources, which generally do not include oil and gas operations for the sources listed above specifically. If regulated at all, oil and gas sources of criteria pollutants and HAPs fall under state minor source programs. The strictest criteria air pollutants regulations apply to areas not meeting established maximum ambient air standards, which are referred to as "non-attainment" areas. Air quality air regulations are, far and away, the most cumbersome regulations related to the oil and gas industry. Unfortunately, sitting in on an air regulation discussion will invariably result in the use of the phrase "it depends" being used more than in pretty much any other discussion one could possibly have. Most confusion related to air regulations stems from the way they are written: they can and often are based on any combination of equipment type, locations, date of manufacture, date of equipment order, date of reconstruction, whether the equipment is considered mobile or not, etc. These regulations are written such that one piece of equipment may be covered by several different regulations, to the point that even the regulatory employees are often as confused as industry personnel.

A review of regulations brings to mind two things: the interaction between Peter Gibbons and "the Bobs" in the movie *Office Space* and the interaction between Abbot and Costello in the equally brilliant skit *Who's on First*?

Peter Gibbons: And here's something else, Bob…I have eight different bosses right now.
Bob Slydell: I beg your pardon?
Peter Gibbons: Eight bosses.
Bob Slydell: Eight?
Peter Gibbons: Eight, Bob. So that means that when I make a mistake, I have eight different people coming by to tell me about it.
Office Space –written and directed by Mike Judge

Abbott: Now, on the St. Louis team we have Who's on first, What's on second, I Don't Know is on third--
Costello: That's what I want to find out. I want you to tell me the names of the fellows on the St. Louis team.
Abbott: I'm telling you. Who's on first, What's on second, I Don't Know is on third--
Costello: You know the fellows' names?
Abbott: Yes.
Costello: Well, then who's playing first?
Abbott: Yes.
Costello: I mean the fellow's name on first base.
Abbott: Who.
Costello: The fellow playin' first base.
Abbott: Who.

Costello: The guy on first base.
Abbott: Who is on first.
Costello: Well, what are you askin' me for?
Abbott: I'm not asking you--I'm telling you. Who is on first.
Costello: I'm asking you--who's on first?
Abbott: That's the man's name.
Costello: That's who's name?
Abbott: Yes.
Who's on First? – performed by Abbott and Costello

Yes, the best way to describe air regulations is to imagine the unholy combination of the two discussions above. Some states cannot, it seems, even maintain constant definitions within their own regulations. For example, in Texas, something as mundane as the definition for "start of construction" varies with differing types of permits. In one type of permit, "start of construction" refers to the beginning of operations, while in others, "start of construction" is completely different, and refers to time before the start of construction of a tank battery.

In an attempt to clear up some of the confusing points of air regulations, the following section is broken into discussions on some common sources and types of air emissions and an overview of a few of the common air regulations and permits related to the oil and gas industry.

12.1 Common Sources of Air Emissions

Air emissions can come from any number of sources, both related to and not related to the oil and gas industry. Non-industry examples are large sources such as refineries, factories, and power plants, smaller sources such as dry cleaners and gas stations, mobile sources such as cars, buses, planes, trucks, and trains, and naturally occurring sources such as natural windblown silt/sand/dust, and even volcanic eruptions. All of these sources contribute to the overall breakdown of air quality. However, keeping in line with the overall context of this book, the discussion in this section will be concentrated on air emissions and regulations related to the oil and gas industry.

Air emissions from shale gas operations can occur at the drill site during drilling and fracturing operations. Air emissions, including diesel fumes, fine particulate matter, and methane, also occur at off-site facilities such as pipelines, natural gas compressors, evaporation pits, and pig launch receiver/launchers. In addition to those emissions linked directly to oil and gas facilities, air emissions are also attributed to trucks used to transport equipment, fracturing fluid ingredients, and water to the wellpad, drilling rigs, compressors, tanks, and pumps.

A principal concern for shale gas emissions is volatile organic compounds (VOC) such as propane, BTEX (benzene, toluene, ethylene, and xylene) constituents, and the six principal criteria pollutants classified by the EPA, nitrogen dioxide (NO2), ozone (O3), sulfur dioxide (SO2), particulate matter (PM), carbon monoxide (CO), and lead (Pb).

12.2 Fugitive Air Emissions

Fugitive air emissions are air emissions that escape from equipment through normal use, routine maintenance, or small leaks. This can be through equipment such as pumps, valves, flanges, compressors, and process drains. There are generally four categories of emissions:

- Normal operations: emissions during expected operating scenarios
- Planned maintenance, start-up and shut-down: emissions during normal operations that are expected and predictable
- Scheduled maintenance start-up and shut-down: unexpected emissions that are expected but not necessarily predictable
- Emission events: emission events that are not authorized, scheduled, or planned

Three sources of emissions commonly related to fracking operations are flare emissions, liquid storage tank emissions, and loading/unloading emissions:

- Flare emissions: flare emissions result from the pilot light, combusted waste, and uncombusted waste (depending on flare efficiency). These can result in the release of NOx, CO, VOCs, SO2, H2S (in some areas), HAPs, CO2, and methane.

- Liquid storage tank emissions: liquid storage tank emissions are categorized as working, breathing, and flash emissions. These can result in the release of VOCs, HAPs, H2S (in some areas), CO2, and methane. Liquid storage tank emissions are most commonly controlled utilizing vapor-recovery units and condensers.
  - Working emissions occur during tank filling and draining
  - Breathing emissions result from normal daily fluctuations in temperature and pressure
  - Flash emissions result from a high pressure stream being directed into an atmospheric tank.
- Loading/Unloading emissions: loading/unloading emissions result from hose disconnects when petroleum liquids are loaded into trucks, rail cars, tankers, or even ships, and can also occur when the vapor space in a tanker truck is displaced by fluids. These can also result in the release of VOCs, HAPs, H2S (in some areas), CO2, and methane, and are normally controlled utilizing vapor recovery units, incineration through flaring or thermal oxidizers, and carbon systems.

In addition to emissions related directly to fracking operations, emissions related to the oil and gas industry also include fugitive emissions of methane from other sources as well, such as pressure relief valves of separators, condensate tanks, and produced water tanks. Controls to prevent and mitigate fugitive emissions are generally comprised of a leak detection and repair program (commonly referred to as LDAR) and maintenance. Emissions from gas production come from:

- Direct emissions (methane venting during frack cleanup, lost gas or fugitive emissions, and CO_2 from fuel combustion)
- Indirect emissions from trucks (pumpers and processing equipment used in drilling, fracturing and production)
- Offsite gas processing equipment such as natural gas in pipelines and compressors from production wells to the point of sale

12.3 Silica Dust Exposure

Silica dust is an emission source that is becoming more of a fracking industry concern. As discussed throughout this book, fracking commonly requires large volumes of sand and water to be pumped into wells at high pressures to break up tight formations, like shale, which have trapped oil and gas. Therefore, many truckloads of sand must be off loaded and transferred before being mixed with water and other chemicals and pumped downhole. The dust produced, which may contain up to 99% crystalline silica, is a health concern due to the risk of silicosis, a progressive and disabling lung disease.

12.4 Stationary Sources

A "stationary source" is, as the name suggests, a non-mobile piece of equipment that emits an air pollutant. Some states have specified time periods, such as six months, that are also used to define a stationary source.

While not specific to fracking operations, common stationary sources in the natural gas industry as a whole include compressor engines, turbines, generators, tanks, reboilers or associated equipment, boilers, vaporizers, flares, etc. Stationary sources may be subject to different requirements depending on the amount and type of air pollutant emitted and the quality of the air in the vicinity of the source.

12.5 The Clean Air Act

The Federal Clean Air Act (Clean Air Act) provides the principal framework for national, state, and local efforts to protect air quality. Under the Clean Air Act, the Office of Air Quality Planning and Standards Organization (OAQPS) is responsible for setting standards known as national ambient air quality standards (NAAQS), for pollutants that are considered harmful to people and the environment.

As with other types of regulations, state and local agencies take the lead in carrying out the requirements of the Clean Air Act. Each state has to develop and implement plans that outline how to control air pollution as required by Clean Air Act. In doing this, state and local governments will need to conduct air monitoring, facility inspections, and conduct permit management and enforcement. Also, as with other types of regulations, individual states may have stronger air quality laws than the federal regulation, but they may not have weaker pollution limits than those set by the EPA.

The Clean Air Act regulates many aspects of the construction and operation of oil and gas facilities, including compressors engines, flares, and other emission sources. Under the Clean Air Act, many stationary sources of air pollution cannot legally operate until an air permit is obtained and compliance requirements are in place and met.

The Clean Air Act has several programs to protect public health and welfare and enhance the quality of the nation's air resources, including New Source Performance Standards (NSPS – found in 40CFR Part 60), National Emission Standards for Hazardous Air Pollutants (NESHAP – found in 40CFR Part 63), NSR, and Prevention of Significant Deterioration (PSD). Depending on the level and source of emissions, various types of air permits may be required.

12.6 Regulated Pollutants

Regulated pollutants that result from emissions at oil and gas locations are generally separated into two main categories:

- National Ambient Air Quality Standard (NAAQS) Criteria Pollutants: there are six NAAQS criteria pollutants that are generally regulated by county and pollutant
- National Emission Standards for Hazardous Air Pollutants (NESHAP): there are 187 HAPS that are regulated on a "source by source" basis, and are considered more difficult to monitor than criteria pollutants

12.7 NAAQS Criteria Pollutants

As discussed earlier, there are six principal criteria pollutants classified by the EPA as criteria pollutants: nitrogen dioxide (NO2), ozone (O3), sulfur dioxide (SO2), particulate matter (PM), carbon monoxide (CO), and lead (Pb). Criteria pollutants are more of an "industry based regulation," or in other words, government regulation of an industry as a whole. Criteria pollutants are also monitored on a "not to exceed" level by county and pollutant requiring regional attainment of established levels.

12.8 Attainment Versus Non-attainment

"Attainment" and "non-attainment" are terms to describe the air quality in a given area (usually broken down by county) for any of the six established criteria pollutants. An area with levels below the established standards for all six criteria pollutant is referred to as an attainment area; a county that does not meet the standards, or is considered to contribute to elevated levels in nearby areas that do not meet the standards, is called a non-attainment area. Industry located in non-attainment areas are held to much more stringent air emission standards than those located in attainment areas. Since an area is compared to established levels for all six criteria pollutants, an area may be designated attainment for some pollutants and nonattainment for others. Once an area is determined to be non-attainment, it is then further broken into tiered levels of non-attainment: marginal, moderate, serious, severe, and extreme. This further classification is determined based on number of criteria pollutants exceeded and levels of exceedance.

In addition to "attainment" and "non-attainment," two more possible classifications for NAAQS criteria pollutants are "unclassified" and "maintenance." Areas that move from non-attainment to attainment retain many of the more stringent rules of attainment areas and are referred to as maintenance areas and areas with insufficient air quality data for classification are referred to as unclassified areas and are treated the same as attainment areas.

12.9 Types of Federal Regulations

There are various types of Clean Air Act regulations and permits. Oil and gas operations, and shale gas in particular, are subject to numerous regulations such as those related to the six criteria pollutants and 187 hazardous air pollutants mentioned earlier. The following is a brief discussion of some of the most common types related to the oil and gas industry.

12.10 MACT/NESHAP HAPs

Unlike the relatively small number of criteria pollutants, there are 187 separate designated hazardous air pollutants – listed pollutants that increase mortality of serious, irreversible, or incapacitating illness. MACT/NESHAP hazardous air pollutant regulations are "risk-based regulation" and are established on "source by source" criteria for each separate piece of equipment; for example, boiler MACT and reciprocating internal combustion engine MACT (NESHAP Quad ZZZZ). Also, since these are risk-based regulations of separate pieces of equipment, there are no established attainment designations of across the board acceptable levels.

12.11 NSPS Regulations: 40 CFR Part 60

In accordance with the Clean Air Act, reconstructed or modified sources are potentially subject to a New Source Performance Standard (NSPS), which is exactly what the name describes, standards set up to manage and determine possible new sources for emissions. If a facility is a new, reconstructed, or modified source, it could be considered a potentially affected source under the NSPS. NSPS standards are in place to create higher standards of performance and maintenance as new equipment is brought into service. New equipment is automatically covered by NSPS and additional sources are added as reconstructed or modified.

A source is "reconstructed" if the components of an existing stationary source are replaced and/or repaired and the fixed capital costs exceed a specified percentage of the cost of constructing a comparable new source. The specified percentage is determined by the applicable air regulation. New-source performance standards may be applied to sources reconstructed after the proposal of the standard if it is technologically and economically feasible to meet the standards.

A source is "modified" if any physical or operational change to an existing facility is made that results in an emission increase of any pollutant to which the standard applies, such as horsepower increase or engine conversion.

Consequently, an existing facility containing an existing (*older*) emission unit, which might not be subject to the NSPS due to the age of the unit, could still become subject to regulation under NSPS if reconstructed or potentially modified.

12.12 NSPS Subpart OOOO

NSPS Subpart OOOO (Quad OOOO) regulations, finalized by the EPA on August 16, 2012, centered on gas production, transmission and distribution, are generally considered the most concentrated on fracking operations. Therefore, this regulation will be discussed in greater detail than those regulations centered on oil and gas as a whole.

12.13 Facilities/Activities Affected by NSPS OOOO

The following is a brief overview of activities/equipment requirements regulated under NSPS OOOO to reduce VOC emissions from sources constructed, reconstructed or modified after August 23, 2011:

- **Natural gas well completions**
  For natural gas wells that use hydraulic fracturing, new regulations require companies to control and reduce emissions during flowback. The goal of the new standard is to reduce VOC emissions generated during flowback activity. In the past, the general industry practice was to vent the gas generated during flowback. Starting October 15, 2012, these new standards require flowback emissions to be routed to a completion combustion device, usually a flare. After January 1, 2015, flowback will be required to implement "green completions" where flowback vapors will be routed to the gas gathering line. If this infeasible due to lack of infrastructure, vapors will be routed to completion combustion device (i.e., flare). The only exceptions to implementing this regulation are if the completion could potentially result in a fire hazard or explosion or if local ordinances restrict the use of flares, such as during a burn ban.
- **Storage vessels**
  EPA regulates storage vessels that handle crude oil, condensate, and produced water at production and gathering/booster stations to reduce emissions by 95%. Storage vessels with VOC emissions of greater than 6 tons per year (tpy) must install emission controls to reduce emissions to less than 6 tpy. Operators must demonstrate compliance by developing emission calculations and include them in annual reports submitted to EPA or the delegated state agency. A storage vessel is exempt, however, if skid mounted (i.e., mobile) and is at the site less than 180 consecutive days.

- **Pneumatic controllers**
  Pneumatic controllers at processing plants and production facilities also have requirements under this new EPA regulation. At the gas processing plants, each continuous bleed natural gas driven pneumatic controller must operate at a zero bleed rate (i.e., instrument air) by October 15, 2012. At production facilities, each continuous bleed natural gas driven pneumatic controller must operate at a bleed rate of 6 standard cubic feet per hour (SCFH) or less. Intermittent bleed and no-bleed controllers are exempt from this regulation.
  The goal of this part of the regulation is to encourage controllers to use low bleed, intermittent-bleed or non-gas-driven devices in the future. Operators can claim exceptions if a high continuous bleed rate controller is required for functional needs, including but not limited to response time, safety, and positive actuation.
- **Compressors**
  The compliance date for both centrifugal and reciprocating compressors was October 15, 2012. For wet seal centrifugal compressors, the regulation stipulates that operators reduce emissions from each centrifugal compressor wet seal fluid degassing system by 95% through routing gas to controls.
  For reciprocating compressors, the new rule requires operators to replace rod packings before the compressor has operated 26,000 hours or prior to 36 months from last replacement or date of startup for the unit, whichever is later.

12.14 Other Types of Federal NSPS and NESHAP/MACT Regulations

In addition to Quad OOOO, which directly relates to affects hydraulic fracturing operations, there are several regulations that affect the oil and gas industry as a whole. The following is a quick discussion of a few of the other regulations that affect industry.

12.15 NSPS Subpart IIII

NSPS Subpart IIII (Quad IIII) establishes regulation of criteria pollutants from compressed ignition combustion engines. This regulation applies to engines operated by diesel or liquefied petroleum gas. It also regulates all sizes of internal combustion engines in emergency and non-emergency use.

12.16 NSPS Subpart JJJJ

NSPS Subpart JJJJ (Quad JJJJ) establishes regulation of criteria pollutants for spark injection internal combustion engines. Quad JJJJ applies to manufacturers, owners, and operators of stationary spark ignition internal combustion engines based on the date the engine was constructed. Engines subject to this rule include natural gas engines and gasoline engines greater than 25hp, including those which are new, reconstructed, and modified.

12.17 NSPS Subpart KKK

NSPS Subpart KKK (Triple K) establishes regulation for equipment at on-shore natural gas processing plants constructed, reconstructed or modified after January 10, 1984 and on or before August 23, 2011. It includes equipment leaks from natural gas processing plants from compressors, pumps, open-ended valves, flanges, and connecters.

12.18 MACT Subpart HH and Subpart HHH

MACT Subpart HH (Double H) establishes emission standards of HAPs related to dehydration units located at oil and natural gas production facilities, while MACT Subpart HHH (Triple H) establishes emission standards of HAPs related to dehydration units located at oil and natural gas transmission and storage facilities.

12.19 MACT Subpart ZZZZ

MACT Subpart ZZZZ (Quad ZZZZ) sets national emission standards and operating limitations for HAP emissions from stationary reciprocating internal combustion and operating limitations for hazardous air pollutants from stationary engines (including compressor station engines) with greater than 500 horsepower put in place prior to June 12, 2006 (Quad JJJJ applies for engines put in place after June 12, 2006).

12.20 Construction and Operating New Source Review Permits

In addition to federal air regulations, there are also several differing types of permits governing new source construction compliance which vary by state. These include construction permits, used to regulate construction and modification of sources and prevent significant deterioration affects to non-attainment areas, and operating permits, authorizing the operation of larger emission sources and to keep these sources more accountable to federal standards.

Prevention of significant deterioration (PSD) is a construction air pollution permitting program designed to ensure that air quality does not degrade beyond the NAAQS levels. The PSD permit includes a requirement to comply with ambient air quality levels and to install "best available control technology" (BACT) for criteria pollutants emitted in significant levels.

12.21 Title V Permits

Title V permits are operating permits that grant larger air emission sources the permission to operate. Title V permits include all air standards and requirements that apply to that source, including emissions limits, monitoring protocol, and record keeping requirements. This permit is used to identify all of the applicable air quality requirements for a site. It also requires that the site demonstrate compliance with the requirements on an ongoing basis, often semi-annually. The main purpose of a Title V permit is intended to be a single point of reference for all regulations at the permitted location. However, along with the intended gathering of all the regulations in one document, Title V permits also include a great deal of additional administrative protocols that more than negate any intended ease of review the document may provide.

13

Chemicals and Products on Locations

A common point of contention for residents in oil producing areas is on-site storage of chemicals and products. For decades, one of the biggest drivers for public concern has been the identity and amounts of chemicals stored on pad locations during all phases of the well completion and production process. This can include fuels used on location, the makeup of drilling fluids (water-based, oil-based or synthetic), fracking chemicals and additives stored on locations, chemicals kept on site during ongoing production, and even the products – produced water and/or oil – stored on site as recovered from the well during operations.

People safely use chemicals everyday: dish soap, gasoline, motor oil, bleach, hydrogen peroxide, and vinegar are some examples. The same chemicals are frequently used in workplaces, but are often in a higher concentration and volume, which could pose a higher risk for individuals. As an example, 3% hydrogen peroxide is used to clean skin abrasions or as a mouthwash at home, and some folks use concentrations as high as 35% for treating swimming pool water.

Companies produce and distribute products with concentrations of hydrogen peroxide at levels between 3% and 35%, so one must be able to recognize which products are generally safe and which products require precautions to be handled and used safely. The important information to take away from this section includes:

- How to recognize that a chemical product may present a hazard
- Know where to get information that explains the hazards
- Know who to ask for help and assistance at any time
- Understand what GHS means and the reason for its implementation
- Understand the difference between GHS and HCS 2012

It is also important that one is able to recognize the changes to hazard communication in the U.S. The most important among these are:

- Label information and changes
- Safety data sheet information and format
- The difference between the numerical rating systems
- Chemical hazards and classification
- Chemical Labels
- Safety Data Sheets (Formerly known as Material Safety Data Sheets)
- Detection of and protection from hazardous chemical exposure

Many people in the chemical industry, as well as those directly or even indirectly related to the industry, should understand what Hazard Communication is. What is Hazard Communication? OSHA's Hazard Communication standard is published in the Code of Federal Regulations at 29 CFR 1910.1200. The standard was first published in 1983. The standard has been revised since then, with the most recent and sweeping changes made in March 2012 to conform to the United Nations Globally Harmonized System of Classification and Labeling of Chemicals (or GHS). Hazard Communication is designed to ensure that employers and employees are equipped with proper information to facilitate safe storage and handling of chemicals.

The general public would be pretty surprised at how little chemical and products are stored on site during construction, drilling, and frack operations. For the most part, oilfield operations have become such a streamlined and efficient operation that operators will know how much of a given chemical product will be necessary and, for the most part, make all attempts to have the chemicals arrive on location as close to when needed as possible to avoid storage. This process is beneficial to the operator in that it cuts down on the time taken up by storing chemicals only to return for them when needed, cuts down on waste from unused or outdated chemicals, cuts down on equipment needed to maneuver chemicals if they are delivered directly to point of need, allows for more working space on the pad, and also helps avoid a great deal of logistical problems related to maneuvering equipment around storage areas.

Once production operations are in place and wells begin producing the fluids – produced water and oil – are often stored on site in large tanks while awaiting transport off-site. Safeguards put in place to protect the environment and public from tank releases include consistent measurements by pumpers, high level shut down sensors, continued equipment observations and maintenance, and secondary containments in place around the tanks to contain any fluids that may release from tank. Secondary containments may be constructed of properly packed and integrity-tested earthen materials or up to specifically designed and manufactured metal containments with plastic liners. No matter the materials used in construction, secondary containments must be sufficiently large enough to contain all the fluids that could possibly escape the tanks plus sufficient extra space for "waste case scenario" rainfall. This amount is calculated for each region of the country based on historic rainfall amounts.

Even with attempts to minimize the amount of on-site storage, some chemical and product storage is unavoidable, and there are very valid concerns, including potential spills, leaks, tank or container overfill, and even the chance of traffic accidents on location or roadways leading to releases of chemicals and/or products. Release events could range from relatively small amounts from equipment leaks to possibly hundreds of barrels from tank release. Two regulatory measures in place to manage and oversee on-site chemical storage conditions are requiring Spill Prevention Countermeasure and Control (SPCC) plans and SARA reporting.

SPCC plans are documents required by all facilities having the potential to discharge oil to navigable waters of the U.S. and meeting one or both of the following: greater than 1,320 gallons (31.4 brls) aggregate aboveground storage in equipment, drums, tanks, totes, tanks greater than 55 gallons in size, or greater than 42,000 gallons total underground storage capacity. Just to clarify, aggregate refers to adding up separate amounts of all storage vessels. There can be one 1,320-gallon tank or ten 132-gallon tanks, and they would be equal under the SPCC requirements. Also, the "having potential to discharge oil to navigable waters of the U.S." is left up to regulatory discretion to calculate, and has come to include pretty much anywhere in the U.S. one could imagine. SPCC plans are, to keep it simple, engineer-stamped documents that must be created for all facilities meeting the above conditions that include a list of spill response procedures, an emergency notification phone list, inspection procedures and schedule, training requirements, site figures, site chemical and product storage vessel types and sizes, and containment calculations to prove sufficient containment is given to contain the largest possible spill amount.

SARA reporting, or possibly better known as the federal "right to know," requires quarterly and annual reporting of chemical storage details (types of chemicals, amounts, and dates of storage) for all facilities that used more than 10,000 pounds per year of the chemical exceeding the threshold quantity. This requirement means a facility storing more uses more than 10,000 pounds of a given chemical in a year must report that chemical and amount. This program is intended as the "right to know" for emergency responders and emergency services that may respond to an emergency situation on the location so they will be able to adequately prepare for what may be stored on site. The drawback of this program as related to the oil and gas industry is that, with quarterly reporting, by the time a chemical has been reported, the oilfield function requiring the chemical has normally been long complete and the chemicals are no longer on site. This basically means that once the chemical is reported as being on a location, it is no longer there; however, as previously stated, oilfield operations have become such a streamlined process that if one knows what has been reported for a previous location by a specified operator one can, for the most part, expect much the same chemicals and products stored at following locations. If one is really curious about all the chemicals used at a site, ask to receive a copy of the Material Safety Data Sheet of the chemicals used.

13.1 Material Safety Data Sheets (MSDS)

Any time a company produces for sale or uses a chemical, a Material Safety Data Sheet (MSDS) has to be written on the product and on file when used. Occupational Safety and Health Administration (OSHA) estimates that there are over 650,000 hazardous chemicals used daily in the United States, and that hundreds more will be added this year alone. To address the physical and health hazards of these chemicals, OSHA finalized the Hazard Communication Standard (HCS) on November 25, 1983. The purpose of the HCS is to "ensure that the hazards of all chemicals produced or imported are evaluated, and that information concerning their hazards is transmitted to employers and employees." (29 CFR 1910.1200(a)(1))

Employers are under obligation to use labels, MSDS, and other information to evaluate both the physical and health hazards created by the use of chemicals in their workplace, establish a program that addresses these hazards, and train workers to minimize their exposure. According to an OSHA Executive Summary: "Chemical information is the foundation of workplace chemical safety programs. Without it, sound management of chemicals cannot occur. The HCS has made provision of hazard information about chemical products an accepted business practice in the United States. There is now a whole generation of employers and employees who have never worked in a situation where information about the chemicals in their workplace is not available."

Manufacturers or importers of chemicals must create or obtain a MSDS for every hazardous chemical that they produce or import (29 CFR 1910.1200(g)), and supply the appropriate one with a customer's first purchase, and any time the MSDS changes (29 CFR 1910.1200(g)(6)(i)). Employers are not required to evaluate information on a MSDS (29 CFR 1910.1200(d)(1)). They do, however, have a duty to study and to use it to "develop, implement and maintain... a written hazard communication program" to ensure the safety of their workers (29 CFR 1910.1200(e)(1)). To help address worker safety "at all times," OSHA requires employers to make MSDS "readily accessible during each work shift to employees when they are in their work area(s)" (29 CFR 1910.1200(g)(8)). OSHA permits electronic and other forms of access to MSDS, as long as there are "no barriers to immediate employee access in each workplace" (29 CFR 1910.1200(g)(8)).

13.2 Contents of an MSDS

Frack site workers, as well as anyone working in an industry or market that uses chemicals, will have access to an MSDS for any chemical with which they may have contact. Interestingly enough, consumer products also have MSDSs. In fact, the local hardware store has a complete file of MSDSs for all the consumer chemicals they sell, and many department stores do as well. If the reader ever wants to know the dangerous effects of a particular insecticide or cleaner, he can refer to the MSDS for detailed information. Beware though; the information contained within an MSDS can be a bit foreboding. Like pharmaceuticals and over-the-counter medications, the warnings typically are meant to take into consideration any and all dangers that may happen upon exposure. Without a working knowledge of the terms and criteria put forth in the MSDS, the layperson could quickly become horrified with the prospect of using a product only to experience dizziness, dry mouth, or shortness of breath (which seems to be the universal response to everything from aspirin to Zoloft®). The following is an explanation that is provided to help interpret the information found on manufacturers' MSDSs. While the format of these data sheets varies from manufacturer to manufacturer, certain components appear on each sheet.

13.3 Product Identification

This section gives the name and address of the manufacturer and an emergency phone number where questions about toxicity and chemical hazards can be directed.

- Product Name: Commercial or marketing name.
- Synonym: Approved chemical name and/or synonyms.
- Chemical Family: Group of chemicals with related physical and chemical properties.
- Formula: Chemical formula, if applicable; i.e., the conventional scientific definition for a material.
- CAS Number: Number assigned to chemicals or materials by the Chemical Abstracts Service.

13.4 Hazardous Ingredients of Mixtures

This section describes the percent composition of the substance, listing chemicals present in the mixture. If it was tested as a mixture, it lists chemicals which contribute to its hazardous nature. Otherwise, it lists ingredients making up more than 1% and all carcinogens.

The OSHA permissible exposure limit (PEL), National Institute for Occupational Safety and Health (NIOSH) recommended exposure limit (REL), and/or the American Conference of Governmental Industrial Hygienists (ACGIH) threshold limit value (TLV) will also be listed, if appropriate.

The OSHA PEL is the regulated standard, while the others are recommended limits. The PEL is usually expressed in parts per million parts of air (ppm) or milligrams of dust or vapor per cubic meter of air (mg/m3). It is usually a time weighted average (TWA), concentration averaged over an eight-hour day. Sometimes, a STEL or short term exposure limit may be listed. The STEL is a 15-minute TWA that should not be exceeded. A ceiling limit, (c), is a concentration that may not be exceeded at any time. A skin notation means that skin exposure is significant in contributing to the overall exposure.

13.5 Physical Data

This section outlines the physical properties of the material. The information may be used to determine conditions for exposure. For example, one can determine whether or not a chemical will form a vapor (vapor pressure), whether this vapor will rise or fall (vapor density), and what the vapor should smell like (appearance and odor). This could help determine whether to use a fume hood or where to place ventilators. The following information is usually included:

- Boiling Point: temperature at which liquid changes to vapor state
- Melting Point: temperature at which a solid begins to change to liquid
- Vapor Pressure: a measure of how volatile a substance is and how quickly it evaporates. For comparison, the VP of water (at 20o C) is 17.5 mm Hg, Vaseline (non-volatile) is close to 0 mm Hg, and diethyl ether (very volatile) is 440 mm Hg.
- Vapor Density (air=1): weight of a gas or vapor compared to weight of an equal volume of air. Density

greater than 1 indicates it is heavier than air, while less than 1 indicates it is lighter than air. Vapors heavier than air can flow along just above ground, where they may pose a fire or explosion hazard.

- Specific Gravity (water=1): ratio of volume weight of material to equal volume weight of water.
- Solubility in Water: percentage of material that will dissolve in water, usually at ambient temperature. Since the much of the human body is made of water, water soluble substances more readily absorb and distribute.
- Appearance/Odor: color, physical state at room temperature, size of particles, consistency, and odor, as compared to common substances. Odor threshold refers to the concentration required in the air before vapors are detected or recognized.
- % Volatile by Volume: Percentage of a liquid or solid, by volume, that evaporates at a temperature of 70°F.
- Evaporation Rate: usually expressed as a time ratio with ethyl ether = 1, unless otherwise specified.
- Viscosity: internal resistance to flow exhibited by a fluid, normally measured in centistokes time or Saybolt Universal Secs.
- Other Pertinent Physical Data: information such as freezing point is given, as appropriate.

13.6 Fire and Explosion Hazard Data

This section includes information regarding the flammability of the material and information for fighting fires involving the material.

- Flashpoint: the lowest temperature at which a liquid gives off enough vapor to ignite when a source of ignition is present.
- Auto-ignition Temperature: the approximate temperature at which a flammable gas-air mixture will ignite without spark or flame. Vapors and gases will

spontaneously ignite at lower temperatures in oxygen than in air.

- Flammable Limits: the lower explosive limit (LEL) and upper explosive limit (UEL) define the range of concentration of a gas or vapor in air at which combustion can occur. For instance, an automobile carburetor controls this mixture: too lean (not enough chemical) or too rich (not enough air, as when you flood your engine), and it will not ignite.
- Extinguishing Media: appropriate extinguishing agent(s) for the material.
- Fire-fighting Procedures: Appropriate equipment and methods are indicated for limiting hazards encountered in fire situations.
- Fire or Explosion Hazards: Hazards and/or conditions that may cause fire or explosions are defined.

13.7 Health Hazard Data

This section defines the medical signs and symptoms that may be encountered with normal exposure or overexposure to this material or its components. Information on the toxicity of the substance may also be presented. Results of animal studies are most often given, i.e., LD50 (mouse) = 250 mg/kg. The terms are usually expressed in weight of chemical per kg of body weight. LD50 or lethal dose 50 is the dose of a substance that will cause the death of half of the experimental animals. LC50 is the concentration of the substance in air that will cause the death of half of the experimental animals. Health hazard information may also distinguish the effects of acute (short-term) and chronic (long-term) exposure.

13.8 Emergency and First Aid Procedures

Based on the toxicity of the product, degree of exposure, and route of contact (eye, skin, inhalation, ingestion, injection), emergency and first aid procedures are recommended in this section. Additional cautionary statements, i.e., note to physician, for first aid procedures, when necessary, will also appear here.

13.9 Reactivity Data

This section includes information regarding the stability of the material and any special storage or use considerations.

- Stability: "unstable" indicates that a chemical may decompose spontaneously under normal temperatures, pressures, and mechanical shocks. Rapid decomposition produces heat and may cause fire or explosion. Conditions to avoid are listed in this section.
- Incompatibility: certain chemicals, when mixed, may create hazardous conditions. Incompatible chemicals should not be stored together.
- Hazardous Decomposition Products: chemical substances that may be created when the chemical decomposes or burns.
- Hazardous Polymerization: rapid polymerization may produce enough heat to cause containers to explode. Conditions to avoid are listed in this section.

13.10 Spill, Leak, and Disposal Procedures

This section outlines general procedures, precautions, and methods for cleanup of spills. Appropriate waste disposal methods are provided for safety and environmental protection.

13.11 Personal Protection Information

This section includes general information about appropriate personal protective equipment for handling this material. Many times, this section of the MSDS is written for large-scale use of the material. Appropriate personal protection may be determined by considering the amount of the material being used and the actual manipulations to be performed.

- Eye Protection: recommendations are dependent upon the irritancy, corrosiveness, and special handling procedures.

- Skin Protection: describes the particular types of protective garments and appropriate glove materials to provide personnel protection.
- Respiratory Protection: appropriate respirators for conditions exceeding the recommended occupational exposure limits.
- Ventilation: air flow schemes (general, local) are listed to limit hazards.

Another important term used recently is GHS. What is GHS? GHS stands for Globally Harmonized System of Classification and Labeling of Chemicals. GHS was conceived in 1992 and adopted by the United Nations Economic Commission for Europe (UNECE) in 2002. GHS is a harmonized classification system for chemical substances and mixtures according to their health, environmental, and physical hazards. The stakeholders for GHS information include workers, employers, consumers, transport workers, and emergency responders. Currently, the regulatory schemes for hazard communication vary between countries and even between agencies in the same country. Because these different systems use different wording, classifications, symbols, shapes, colors, etc., there are many opportunities that can, and do, lead to confusion and even trade barriers. GHS is an attempt to reduce the differences, especially those between countries, so information on the hazards of chemicals is transmitted more efficiently.

Hazard Communication Standard 2012 (or HCS 2012) is OSHA's revised Hazard Communications standard. OSHA published the final rule for the modified HazCom standard on March 26th, 2012. It went into effect on May 25th, 2012 with a multi-year phase-in schedule. This modification is intended to conform the standard's requirements with the standard language and methodologies established by the GHS. The major modifications to the Hazard Communication standard include the following:

- Revised criteria for classification of chemical hazards
- Revised labeling provisions that include requirements for use
- Standardized signal words, pictograms, hazard statements, and precautionary statements

- A specified format for safety data sheets
- Related revisions to definitions of terms used in the standard, and requirements for employee training on labels and safety data sheets.

By June 1, 2015, chemical manufacturers and importers must comply with all the requirements of the new rule.

By December 1, 2015, chemical distributors must include compliant labels on all shipments of chemical containers.

By June 1, 2016, all employers that use, handle, or store chemicals must update their workplace hazard communication program as necessary, and provide additional employee training for newly identified physical or health hazards.

The revised hazard communications standard (HCS 2012) requires chemical manufacturers and importers to classify their chemicals according to the hazards laid out in the standard. This is a change from the hazard determinations/evaluations performed under the previous HazCom standard.

First, a chemical is hazardous if it can be classified as a

- Physical hazard
- A health hazard
- An environmental hazard, or simple asphyxiant
- Combustible dust
- Pyrophoric gas
- Or hazard not otherwise classified (HNOC)

A hazard class is a subset of a chemical hazard. Hazard class is the nature of the physical or health hazards. For example: a flammable solid is a class of physical hazard.

A hazard category is a subcategory of a hazard class. Each hazard class has 1 to 4 categories to denote the division of severity within each hazard class. For example: oral acute toxicity is broken up into 4 different categories, with 1 being the most hazardous, and 4 being the least hazardous. Hazard categories compare hazard severity within a hazard class.

A chemical is considered a physical hazard if it is classified as possessing one of the following hazards:

- Explosives: An explosive chemical is a solid or liquid chemical that is in itself capable by chemical reaction of producing gas at such a temperature and pressure and at such a speed as to cause damage to the surroundings. Pyrotechnic chemicals are included even when they do not evolve gases.

- Flammable gases: Flammable gas means a gas having a flammable range with air at 20°C (68°F) and a standard pressure of 101.3 kPa (14.7 psi).
- Flammable aerosols: Aerosol means any non-refillable receptacle containing a gas compressed, liquefied, or dissolved under pressure, and fitted with a release device allowing the contents to be ejected as particles in suspension in a gas, or as a foam, paste, powder, liquid, or gas.
- Oxidizing gases: Oxidizing gas means any gas that may, generally by providing oxygen, cause or contribute to the combustion of other material more than air does.
- Gases under pressure: Gases under pressure are gases that are contained in a receptacle at a pressure of 200 kPa (29 psi) or more, or that are liquefied or liquefied and refrigerated. They are comprised of compressed gases, liquefied gases, dissolved gases, and refrigerated liquefied gases.
- Flammable liquids: Flammable liquid means a liquid having a flash point of not more than 93°C (199.4°F). Flash point means the minimum temperature at which a liquid gives off vapor in sufficient concentration to form an ignitable mixture with air near the surface of the liquid.
- Flammable solids: Flammable solid means a solid that is a readily combustible solid, or that may cause or contribute to fire through friction. Readily combustible solids are powdered, granular, or pasty chemicals that are dangerous if they can be easily ignited by brief contact with an ignition source, such as a burning match, and if the flame spreads rapidly.
- Self-reactive substances: Self-reactive chemicals are thermally unstable liquid or solid chemicals liable to undergo a strongly exothermic decomposition even without participation of oxygen (air). This definition excludes chemicals classified under this section as explosives, organic peroxides, oxidizing liquids, or oxidizing solids. A self-reactive chemical is regarded as possessing explosive properties when in laboratory

testing the formulation is liable to detonate, to deflagrate rapidly, or to show a violent effect when heated under confinement.

- Pyrophoric liquids: Pyrophoric liquid means a liquid that, even in small quantities, is liable to ignite within five minutes after coming into contact with air.
- Pyrophoric solids: Pyrophoric solid means a solid that, even in small quantities, is liable to ignite within five minutes after coming into contact with air.
- Self-heating substances and mixtures: A self-heating chemical is a solid or liquid chemical, other than a pyrophoric liquid or solid, that, by reaction with air and without energy supply, is liable to self-heat; this chemical differs from a pyrophoric liquid or solid in that it will ignite only when in large amounts.
- Substances and mixtures that in contact with water emit flammable gases: Chemicals that, in contact with water, emit flammable gases are solid or liquid chemicals that, by interaction with water, are liable to become spontaneously flammable or to give off flammable gases in dangerous quantities.
- Oxidizing liquids: Oxidizing liquid means a liquid that, while in itself not necessarily combustible, may, generally by yielding oxygen, cause, or contribute to, the combustion of other material.
- Oxidizing solids: Oxidizing solid means a solid that, while in itself is not necessarily combustible, may, generally by yielding oxygen, cause, or contribute to, the combustion of other material.
- Organic peroxides: Organic peroxide means a liquid or solid organic chemical that contains a double oxygen structure – it is considered a derivative of hydrogen peroxide. The term organic peroxide also includes organic peroxide mixtures containing at least one organic peroxide. Organic peroxides are thermally unstable chemicals that may undergo exothermic self-accelerating decomposition.

In addition, it may have one or more of the following properties:

- Be liable to explosive decomposition;
- Burn rapidly;

- Be sensitive to impact or friction;
- React dangerously with other substances; or
- Corrosive to metals: A chemical that is corrosive to metals means a chemical which by chemical action will materially damage, or even destroy, metals
- A chemical is considered a heath hazards class if it is classified as one of the following hazards:
  - Acute toxicity: Acute toxicity refers to those adverse effects occurring following oral or dermal administration of a single dose of a substance, or multiple doses given within 24 hours, or an inhalation exposure of 4 hours.
  - Skin corrosion or irritation
  - Skin irritation is the production of reversible damage to the skin following the application of a test substance for up to 4 hours.
  - Skin corrosion is the production of irreversible damage to the skin; namely, visible necrosis through the epidermis and into the dermis, following the application of a test substance for up to 4 hours.
  - Serious eye damage and eye irritation
  - Serious eye damage is the production of tissue damage in the eye, or serious physical decay of vision, following application of a test substance to the anterior surface of the eye, that is not fully reversible within 21 days of application.
  - Eye irritation is the production of changes in the eye following the application of test substance to the anterior surface of the eye that is fully reversible within 21 days of application.
  - Respiratory or skin sensitization
  - Respiratory sensitizer means a chemical that will lead to hypersensitivity of the airways following inhalation of the chemical.
  - Skin sensitizer means a chemical that will lead to an allergic response following skin contact.
  - Germ cell mutagen (cause genetic defects): Mutation is defined as a permanent change in the amount or structure of the genetic material in a cell. The term mutation applies both to heritable genetic changes and to the underlying DNA modifications when known. The term mutagenic and mutagen

will be used for agents giving rise to an increased occurrence of mutations in populations of cells and/or organisms.

- Carcinogen (cancer causing agent): Carcinogen means a substance or a mixture of substances that induce cancer or increase its incidence. Substances and mixtures that have induced benign and malignant tumors in well-performed experimental studies on animals are considered also to be presumed or suspected human carcinogens unless there is strong evidence that the mechanism of tumor formation is not relevant for humans.
- Reproductive toxin: Reproductive toxicity includes adverse effects on sexual function and fertility in adult males and females, as well as adverse effects on development of the offspring. Some reproductive toxic effects cannot be clearly assigned to either impairment of sexual function and fertility or to developmental toxicity. Nonetheless, chemicals with these effects shall be classified as reproductive toxicants.
- Specific target organ systemic toxicity: Single and repeated exposures that are toxic to specific organ(s).
- Specific target organ toxicity: single exposure (STOT-SE) means specific, non-lethal target organ toxicity arising from a single exposure to a chemical. All significant health effects that can impair function, both reversible and irreversible, immediate and/or delayed.
- Specific target organ toxicity: repeated exposure (STOT-RE) means specific target organ toxicity arising from repeated exposure to a substance or mixture. All significant health effects that can impair function, both reversible and irreversible, immediate and/or delayed.
- Aspiration hazard (chemical pneumonia): Aspiration means the entry of a liquid or solid chemical directly through the oral or nasal cavity,

or indirectly from vomiting, into the trachea and lower respiratory system. Aspiration toxicity includes severe acute effects such as chemical pneumonia, varying degrees of pulmonary injury, or death following aspiration.

The effects of health hazards are normally classified as acute or chronic. Acute hazards are those whose effects occur rapidly after exposure. Example: a skin corrosive will begin to burn the skin immediately upon contact. Chronic hazards are those whose effects take prolonged or repeated exposure.

Example: repeated exposure to benzene over several years has been shown to increase the risk of cancer. For hazardous chemicals that do not meet the specific criteria for a physical or health hazard classes that were previously shown, a separate classification, Hazards Not Otherwise Classified (HNOC) was created. HNOC classification is available for chemicals whose hazards fall below an assigned or cut-off value or concentration limit of the hazard class or provides a place for assignment to a GHS hazard category that has not been adopted by OSHA.

Labels are required on almost all containers of hazardous chemicals in the work place, including well sites. The label is the most immediate source of information on hazardous chemicals, so it is important to understand what information can be found on the label that will help you to safely use the chemical. Under the new Hazard Communication guidelines, most hazardous chemical labels must include the following information:

- Product Identifier
- Pictograms
- Signal words
- Hazard Statements
- Precautionary Information
- Supplier Identification

Under the new requirements, pictograms are required on most chemical labels.

Pictograms are to be black symbols on a white background within a red frame.

(**a**) GHS Explosive hazard Pictogram	(**b**) GHS Flammable hazard Pictogram	(**c**)GHS oxidizer hazard Pictogram

(**d**) GHS Toxic hazard Pictogram	(**e**) GHS Health hazard Pictogram	(**f**) GHS Corrosive hazard Pictogram

(**g**) GHS Exclamation point Hazard pictogram	(**h**) GHS Environment Hazard pictogram	(**i**) GHS Compressed gas Hazard pictogram

Signal words and hazard statements will now be included in Labels and Safety Data Sheets. Signal words will be used to emphasize and distinguish hazard levels.

The word "CAUTION" will no longer be used as a signal word. Only two words will be used as signal words.

WARNING: used for less severe hazards

DANGER: used for more severe hazards

In addition, each hazard classification and category will have a specific hazard statement.

Here are four examples from the Flammable classification with the appropriate hazard statement for each category.

Category 1: Extremely flammable liquid and vapor
Category 2: Highly flammable liquid and vapor
Category 3: Flammable liquid and vapor
Category 4: Combustible liquid

Labels may also contain information from other hazard ranking systems.

These rating systems are widely used and commonly found on hazardous chemical labels.

The severity of these hazards are listed on a scale of 0 – 4, with 0 being least hazardous and 4 being most hazardous. Note that this scale is the opposite order of hazard potential used by OSHA. The HMIS label is similar to NFPA in that it has color-coded boxes for hazardous chemicals. In this system:

- Blue Square: Health hazards
- Red Square: Flammability hazards
- Yellow Square: Reactivity hazards
- White Square: Personal protection needed when handling

HMIS also rates the hazards on a scale of 0 - 4 (0 being least hazardous, 4 being most hazardous), similar to NFPA, and is also the opposite order of hazard potential selected by OSHA. GHS uses categories for each hazard class to rate the hazard severity. These categories are typically divided using a scale from 1-4 where 1 is the most hazardous and 4 is the least hazardous. The severity rating is in opposite order of the existing NFPA/HMIS hazard rating system that uses categories 0-4. In these systems, "0" is the least hazardous and "4" is the most hazardous. Since the HCS 2012 hazard rating does not replace NFPA/HMIS hazard ratings, you may see hazardous chemical container labels and SDS with OSHA and NFPA/ HMIS ratings. Since the rating systems are opposite in scale, it is important to understand how both rating scales work as to not be confused.

Labels are also required to include precautionary statements to provide instructions on proper handling of the chemical. There are four types of precautionary statements presented

- **Prevention:** Protective equipment, safe handling and use
- **Response:** First aid and spill cleanup
- **Storage:** Proper storage procedures
- **Disposal:** Proper disposal procedures

A HCS 2012 label should contain:

- The product identifier
- The supplier information
- Precautionary Statements
- Signal Word
- Hazard Pictograms
- Hazard Statements and
- Other supplemental information for the user.

It is important to note that while HCS 2012 is phased in over the next few years,

one may still encounter labels meeting previous HazCom rules. They may not have the same amount of information or pictograms as with new labels, but will still require certain safety information:

- Identity of hazardous chemical
- Appropriate hazard warnings
- Name and address of manufacturer, importer, or other responsible party

Many hazardous chemical labels are complaint with the previous Hazard Communication Standard. Hazardous chemicals in the workplace must be labeled with important information that one needs to know. Each product container used shall be labeled with the product name including:

- Major hazards using words and pictograms
- Precautionary information for protection against hazards
- Name of the manufacturer
- Contact information for the manufacturer

13.12 HCS 2012 Safety Data Sheets (SDS)

Like the MSDS with which the reader is familiar, the Safety Data Sheet (SDS) shall be available for every hazardous chemical product. The SDS (formerly MSDS or Material Safety Data Sheet) is intended to provide comprehensive safety information about a chemical substance or mixture. The revised hazard communications standard now requires a standardized format for all safety data sheets.

Changes to Safety Data Sheets include the following:

- Name changed from Material Safety Data Sheets (MSDS) to Safety Data Sheets (SDS),
- Require the information to be placed in a standard 16 section format,
- Use of the new Hazard classifications just discussed,
- Symbol(s), or a description of symbol(s), in section 2,
- A Signal word (Warning or Danger, as previously discussed),
- Hazard and Precautionary (H&P) statements, and
- The percentages of each hazardous ingredients in the product unless that data is classified as a trade secret

OSHA now requires that the SDS adhere to a standard 16 section format.

Section 1, Identification includes the product identifier or name; manufacturer or distributor name, address, phone number; emergency phone number; recommended use; and restrictions on use.

Section 2, Hazard(s) identification includes all covered hazards regarding the chemical; required label elements.

Section 3, Composition/information on ingredients includes information on chemical ingredients; trade secret claims.

Section 4, First-aid measures includes important symptoms/ effects, acute, delayed; required treatment.

Section 5, Fire-fighting measures lists suitable extinguishing techniques, equipment; chemical hazards from fire.

Section 6, Accidental release measures lists emergency procedures; protective equipment; proper methods of containment and cleanup.

Section 7, Handling and storage lists precautions for safe handling and storage, including incompatibilities.

Section 8, Exposure controls/personal protection lists OSHA's Permissible Exposure Limits (PELs); Threshold Limit Values (TLVs); appropriate engineering controls; personal protective equipment (PPE).

Section 9, Physical and chemical properties lists the chemical's characteristics such as the following:

- Chemical appearance
- Odor, odor threshold

- Physical properties (pH, flash point, flammability limits, vapor pressure and density, auto-ignition temperature, decomposition temperature, viscosity, etc.)

Section 10, Stability and reactivity lists chemical stability, hazardous decomposition products, and conditions to avoid.
Section 11, Toxicological information includes routes of exposure; acute and chronic effects; toxicity data; carcinogenicity.
Section 12, Ecological information

- Ecotoxicity
- Biodegradability

Section 13, Disposal considerations

- Waste description
- Waste handling and disposal

Section 14, Transport information

- DOT shipping name, hazard class, packing group
- UN number

Section 15, Regulatory information lists product specific safety, health, and environmental regulations.
Section 16, Other Information includes the date of preparation or last revision.
Sections 12, 13, 14 and 15 have been allotted to contain information not regulated by OSHA, but agencies, such as EPA, may regulate this information. As a result, OSHA will not enforce compliance on these sections, so these sections may be found blank.

It is important to note that while HCS-2012 is phased in over the next few years, employees will still encounter MSDS with information meeting previous HazCom rules.

For this reason, it will be important to understand the information in both an MSDS and SDS. The difference between the two by looking at what each is required to contain.

MSDS are required to include certain information for the hazardous chemical:

- Chemical ID
- Physical and chemical characteristics

- Physical and health hazards
- Primary routes of entry
- Applicable exposure limits
- Handling, disposal, and spill procedures
- Control measures (Engineering, administrative, PPE)
- First aid
- Preparation date
- Name and contact information of manufacturer

A Safety Data Sheet (SDS) will contain the following information:

- Product specific information on the hazards of the product
- Measures that one can take to protect oneself, others, and the environment,
- Guidelines for proper storage, transportation and disposal procedures,
- Manufacturer contact information, and
- Guidelines on emergency response measures that may be taken by trained and equipped personnel

Before using any hazardous chemical, you should read the SDS and be come familiar with properties of the chemical such as:

Appearance and odor

- Odor is normally the first indication on whether one is being exposed to a chemical.
- For many hazardous chemicals, if one can smell it, one is already overexposed.

Physical Properties: Is it a liquid or solid? Will it give off a large amount of vapor?

Chemical Hazards: Unknown chemicals should be considered hazardous, until otherwise specified. Personal protective equipment (or PPE) is used when hazards may not be completely controlled or eliminated by other control measures (e.g., ventilation).

Types of PPE normally used include

- Chemical-resistant goggles
- Face shield
- Chemical gloves
- Apron
- Respirator

PPE should be checked prior to use. Damaged or malfunctioning PPE must not be used.

PPE selection is based on hazard assessment and effectiveness of PPE.

The SDS should also contain basic first aid measures to be taken in the event that a person is exposed to the hazardous chemical. These are actions that can be taken to reduce the damage that the exposure may cause. Here are some examples actions that may be recommended on the SDS:

- If hazardous chemicals come into contact with the skin or eyes from spills or splashes: flush the affected area immediately with water for at least 15 minutes.
- If a large amount of chemical is inhaled: move away from the source to an area of fresh air.
- If a hazardous chemical is ingested: contact a medical service provider such as a Poison Control Center, Emergency Room, or other Emergency Service provider immediately. Remember, for any hazardous chemical exposure, seek medical attention as soon as possible.

In emergencies when there is no other help available, one can call CHEMTREC (1-800-424-9300). Many hazardous chemical manufacturers include this number on their label and SDS. CHEMTREC is available 24/7/365 days a year. CHEMTREC can provide SDS information on many products and information to Emergency Responders.

14

Public Perception, the Media, and the Facts

In a report published by Energy Institute *Fact-Based Regulation for Environmental Protection in Shale Gas Development* by Charles G. Groat, Ph.D.,

Principal Investigator and Thomas W. Grimshaw, Ph.D., Co-Principal Investigator of the University of Texas at Austin), researchers analyzed print, broadcast, and online news media coverage of shale gas development in the Marcellus, Haynesville, and Barnett shale areas. They found that the tone of media coverage has been overwhelmingly negative in all forms of media. Roughly two-thirds of the articles and stories examined were deemed negative, a finding that was consistent nationally and at local levels.

Researchers also found that less than 20% of newspaper articles on hydraulic fracturing mention scientific research related to the issue. Similarly, only 25% of broadcast news stories examined made reference to scientific studies, and about 33% of online news coverage mentioned scientific research on the issue.

Table 14.1 Tone of media coverage.

Media Coverage	Negative	Neutral	Positive
National Newspapers	64%	25%	12%
Local Newspapers	65%	23%	12%
National Television & Radio	64%	19%	18%
Local Television	70%	27%	3%
Online News	63%	30%	7%

Hydraulic fracturing of shale formations to extract natural gas has no direct connection to reports of groundwater contamination, based on evidence reviewed in a study released by the Energy Institute at The University of Texas at Austin. The study, released at the annual meeting of The American Association for the Advancement of Science (AAAS) in Vancouver, British Columbia, found that many problems ascribed to hydraulic fracturing are related to processes common to all oil and gas drilling operations, such as casing failures or poor cement jobs.

University researchers also concluded that many reports of contamination can be traced to above-ground spills or other mishandling of wastewater produced from shale gas drilling, rather than from hydraulic fracturing. Charles "Chip" Groat, one of the authors of the study and the Energy Institute associate director said "These problems are not unique to hydraulic fracturing."

The controversies surrounding shale gas development have received considerable media coverage. Public perceptions have been influenced by the controversies and media coverage. For these reasons, both media coverage and public perception of shale gas development have been investigated. All three shale gas areas were assessed for media coverage. Public perception was determined for the Barnett Shale area.

Media coverage of hydraulic fracturing, a critical and distinctive component of shale gas development, was assessed for tonality (negative or positive) and reference to scientific research. The assessment covered the period from June 2010 to June 2011 and included three areas:

- Barnett shale area (Dallas, Tarrant, and Denton counties, Texas)
- Haynesville shale area (Shreveport, Louisiana)
- Marcellus shale area (six states)

The six Marcellus locations were Pennsylvania (Pittsburgh), New York (Buffalo), West Virginia (Charleston), Maryland (Hagerstown), Ohio (Cleveland), and Virginia (Roanoke).

Four types of media – newspapers (national and metropolitan), television (national and local), radio (national and local), and online (Google News) – were included using searches for keywords for hydraulic fracturing in 14 groups as follows:

- Well Blowout Pipeline Leaks
- Water Well Contamination Regulatory Enforcement
- Frac Fluid (Frack Fluid) Local Government Response
- Surface Spills or Accidental Release Public Interest and Protest Groups
- Flow-Back Water Barnett Shale Groups
- Water Disposal Wells Wyoming Groups
- Atmospheric Emissions and Air Quality Marcellus Group

Media coverage of shale gas development was assessed in the Marcellus, Haynesville, and Barnett shale areas. The analysis of the tonality of articles and broadcasts included 13 newspapers (three national and 10 metropolitan), 26 broadcast media (seven national and 18 metropolitan television stations and one national radio station), and one online news source.

For the nation as a whole, the attitudes were found to be uniformly about two-thirds negative.

	Negative	Neutral	Positive
National Newspapers (3)	64%	25%	12%
Metropolitan Newspapers (10)	65%	23%	12%
National Television & Radio (7)	64%	19%	18%
Metropolitan Television (18)	70%	27%	3%
Online News (1)	63%	30%	7%

The local media coverage for each of the shale areas shows similarity to the national results for the Barnett and Marcellus shale areas; the Haynesville area may be anomalous because only one newspaper and one television source were available.

	Negative	Neutral	Positive
Barnett Shale Area			
Newspapers (3)	79%	6%	16%
Television (6)	70%	30%	0%
Marcellus Shale Area			
Newspapers (6)	67%	25%	8%
Television (11)	74%	20%	6%
Haynesville Shale Area			
Newspapers (1)	8%	46%	46%
Television (1)	0%	100%	0%

With respect to reference to scientific research, the search found that few articles referenced research on the topic of hydraulic fracturing:

Percent Referencing Newspaper Articles
Scientific Research 18%
Television Reports 25%
Radio Coverage 15%
Online Coverage 33%

Public perception of hydraulic fracturing was assessed specifically in the Barnett shale area, utilizing an online survey method that included 75 questions in six categories:

- Thoughts about hydraulic fracturing
- Perceptions about hydraulic fracturing
- Knowledge of hydraulic fracturing behaviors
- Media use demographics

The area included was expanded to 26 counties in Texas, and the survey included nearly 1500 respondents. The results of the survey indicate a generally positive attitude toward hydraulic fracturing, with more favorable responses for the following descriptors: good for the economy, important for U.S. energy security useful, important, effective, valuable, and productive. Attitudes were neutral to slightly positive as indicated by response to several descriptors for hydraulic fracturing: importance for U.S. energy security, safety, beneficial or good, wise, and helpful.

There was a more negative attitude, however, about environmental concerns. Hydraulic fracturing was felt to be bad for the environment by about 40% of the respondents. Another 44% were neutral, and only 16% were positive. With respect to knowledge of hydraulic fracturing, many respondents were found to have some general knowledge about the process of hydraulic fracturing, but they tend to lack an understanding of regulation and the cost-benefit relationship of production:

- Most respondents overestimate the level of hydraulic fracturing regulation; for example, 71% were not aware that the Railroad Commission does not regulate how close a gas well can be drilled to a residential property.
- Many respondents (76%) overestimate annual water consumption for shale gas usage and underestimate (75%) the amount of electricity generated from natural gas.
- Most generally understand the process of fracturing and gas development surrounding the fracturing of wells, but the scope and technical aspects of fracturing are less well understood. For example, 49% were unaware of proppants, and 42% overestimated scientific evidence surrounding the issue of hydraulic fracturing and water contamination.

Knowledge of policy issues related to groundwater contamination, such as the disclosure of chemicals used in fracturing and active groups affiliated with groundwater issues, was high.

- Knowledge of the occurrence of well blowouts in hydraulic fracturing was high (73%), as well as the impact of blowouts comparison to surface spills (72%). 54% understand the frequency that blowouts have occurred in the Barnett shale.

Hydraulic fracturing knowledge was also assessed for the following five areas:

1. Awareness of Hydraulic Fracturing: 50% of the respondents consider themselves to be somewhat aware or very aware hydraulic fracturing. The other 50% were not very aware or were not aware at all.
2. Concern about Water Quality: 35% indicated they were very concerned, and 40% were somewhat concerned, 24% were not very concerned or not at all concerned.
3. Disclosure of Chemicals Used in Hydraulic Fracturing: Regarding whether state and national officials are doing enough to require disclosure, 12% thought that the officials are doing everything they should, and 32% indicated that officials were doing some of what they should. 47% indicated not as much as should be done was being done. 9% thought that nothing at all was being done.
4. Message to Politicians: When asked about relative priorities of energy production on the one hand and public health and the environment on the other, 67% indicated higher priority on public health and the environment.
5. America's Future Energy Production: The survey also included an assessment of the degree of willingness to get involved in community efforts, such as organizing, protesting, calling legislators, and petitioning.

The results indicate that people are either undecided or ambivalent, or they sense two equal points of view and are not sure which one to accept. It also appears that respondents sense that it is not desirable to get involved: they are mostly unwilling to participate in any events in support of or against hydraulic fracturing. This could be related to their ambiguous attitudes.

14.1 Regulation or Policy Topics: Media Coverage and Public Perception

- Media coverage in the Barnett, Haynesville, and Marcellus shale areas is overwhelmingly negative, with about two-thirds of coverage – including all media and all shale areas – on the negative side.
- Much of the coverage has focused on environmental issues such as groundwater contamination.
- Most residents understand the general process of hydraulic fracturing but lack a strong comprehension of its cope and technical aspects, such as the depth of wells and the role of proppants.
- Reference to shale gas research is understated in the media, with only 15 to 30% of articles and reports containing such references.
- Public perception in the Barnett shale area is somewhat more balanced on the positive and negative side, but a view is held by many (40%) that hydraulic fracturing is bad for the environment. Only 16% have a favorable view.
- Residents in the Barnett shale area are generally informed about hydraulic fracturing, but they tend to overestimate existing regulations and the amount of water used and under estimate the importance of natural gas in electric power production.
- Negative media reporting and public perceptions must be addressed by both regulators and the shale gas industry as regulations are developed or added.
- Comparison of the assessment of public knowledge of hydraulic fracturing in the Barnett shale area with a national survey found that more residents perceived themselves to be more aware of hydraulic fracturing than is the case nationally. But the Barnett area residents and national survey populations are similar in their concerns about water quality and what politicians are doing about the issue.

Table 14.2 Fracking Incidents.

Well Location	When	Incident and Company Involved	Assessment of Incident	Is Underground Hydraulic Fracturing a Direct Cause?	Are Other Drilling Practices at Issue?
Dimock Township, north central PA	2009/2010	**CABOT OIL & GAS.** Issues began with a water well explosion. Gas escaped into aquifer and built up until pressure caused explosion. Residents and national environmental groups alleged Cabot drilling contaminated this and other wells and complained of foul smells and flammable water. Cabot said deep-underground fracking could not have been the cause and on this point PA regulators agree: they told us it was "not Marcellus gas." Cabot provided water to residents involved and litigation is ongoing. Attracted intense national media coverage and attention from environmentalists.	PA regulators blamed Cabot's well design and cement job for allowing naturally-occurring shallow gas to migrate into water supplies of 14 homes; it pointed out last year that despite finding well problems, "hydro fracturing activity has not impacted local wells." It forced Cabot to plug three wells in April, fined the company, barred it from drilling new wells in Dimock for a year, and criticized its slow response. Cabot acknowledges it did not test water wells for pre-existing gas (common in this region). It says it believes its operations did not cause the gas migration, and subsequent tests show that a majority of area wells contain measurable quantities of naturally occurring gas.	NO	YES

Well Location	When	Incident and Company Involved	Assessment of Incident	Is Underground Hydraulic Fracturing a Direct Cause?	Are Other Drilling Practices at Issue?
Caddo Parish, near Shreveport, LA	2010	**EXCO RESOURCES.** EXCO Resources contacted authorities and over 100 homes were temporarily evacuated at company expense in April when the company struck a layer of gas thousands of feet above the Haynesville shale and it escaped into the air and bubbled up through the ground. EXCO had not yet done any fracking of the well.	LA regulators worked with company to investigate cause of gas migration; officials told us they believe a cement job from an adjacent well is at fault. EXCO has agreed to plug 2 wells and says it is continuing to test the water; it says it will incorporate lessons learned into new well designs.	NO	YES
Bainbridge Township, Geauga County, OH	2007	**OHIO VALLEY ENERGY SYSTEMS CORP.** Natural gas seeped into an aquifer and led to an explosion in the basement of a home. The discovery of gas in water supplies drew allegations that fracing by Ohio Valley Energy Systems Corp. had pushed gas to the surface. Residents were not injured and the company worked with regulators to evacuate and house the displaced residents, and stop gas flow.	OH regulators concluded in a lengthy report that the cause was a defective cement job in the well casing, compounded by operator error. The investigation found no evidence of hazardous drilling chemicals in the wells and said the problem would have occurred even if the well had never been hydraulically fractured.	NO	YES

(Continued)

Table 14.2 (*cont.*)

Well Location	When	Incident and Company Involved	Assessment of Incident	Is Underground Hydraulic Fracturing a Direct Cause?	Are Other Drilling Practices at Issue?
Clearfield County, PA	2010	**EOG RESOURCES.** A June 3 blow-out of a gas well sent gas and at least 35,000 gallons of drilling wastewater into the sky and over the ground for 16 hours. This incident occurred during the post-frack flowback period. The incident occurred in a rural, relatively unpopulated area.	PA regulators temporarily suspended EOG's drilling and fracking activities statewide until it investigated the cause and have allowed drilling to resume. EOG said its preliminary assessment was that the seal integrity between the pipe rams of a blow- out preventer and tubing was compromised.	YES – CONTROL OF FLOWBACK WASTE	YES
Garfield County, CO	2001 and 2004	**ENCANA.** A resident, Laura Amos, alleged her water well was contaminated by frack fluids from wells near her house and that the fluids caused an adrenal tumor. Her case has been extensively publicized and she has appeared in a number of environmental groups' reports. Fracking operations near her home occurred 2,000 feet underground and her well is 225 feet deep. (This depth is considerably shallower than shale wells in PA, LA, TX, and AK.)	CO regulators tested her water repeatedly and did not find contaminants associated with frack fluids, including benzene. Encana denied fracking contaminated her water, but Colorado regulators found it in violation of state rules preventing freshwater contamination by gas. Amos has settled with Encana. A study has found correlation of oil and gas drilling with the country's water characteristics.	NO	UNCLEAR

Well Location	When	Incident and Company Involved	Assessment of Incident	Is Underground Hydraulic Fracturing a Direct Cause?	Are Other Drilling Practices at Issue?
Pavillion, WY	Past 10 years	**ENCANA.** EPA tested wells in an area where residents have complained over a decade about effects of gas drilling on their water. In 2009, EPA said it had found chemicals that environmental groups allege are used in the hydraulic fracturing process. EPA says the chemicals "might not be attributable to well components" and also noted agricultural activity nearby.	EPA cautions it does not yet know if there is an oil and gas link and that it will release further study results in August 2010. Encana told us the chemicals at issue are not used in fracking and it needs to see additional results before commenting further.	ALLEGED, NOT PROVEN	UNCLEAR
Caddo Parish, near Shreveport, LA	2009	**CHESAPEAKE ENERGY.** Seventeen cattle were found dead near a Chesapeake Energy drilling site. Chesapeake said Schlumberger was the service company on the job. Both companies denied wrongdoing. The incident attracted extensive coverage; one company not involved told us that when it happened, "nobody wanted to talk about anything else" when it called on state and local officials. Witnesses reported hearing cows bellow before they fell over dead.	LA regulators concluded fluid leaked from a well pad and ran into an adjacent pasture. It fined each company $22,000. Chesapeake says after testing that the cause of death to cattle was inconclusive. Chesapeake and Schlumberger say they have taken a leading role in "enhancing the standard" for well site construction and liquids handling.	NO, BUT HANDLING OF FLUID ABOVE GROUND AT ISSUE	YES

(Continued)

Table 14.2 (*cont.*)

Well Location	When	Incident and Company Involved	Assessment of Incident	Is Underground Hydraulic Fracturing a Direct Cause?	Are Other Drilling Practices at Issue?
	2009	**COMPANY LINK UNCLEAR.** A fish kill along a 43-mile span of the creek due to an invasive saltwater species of golden algae was tied by a number of organizations to hydraulic fracturing. The algae thrive in salty water, and discharge of shale well "flowback water" was suspected because it has high salt content. A gas drilling organization argued drilling activity had not taken place near the relevant portion of the stream. The fish kill continues to be a heated topic among fly-fishing and outdoor enthusiasts.	An interim EPA report blaming golden algae for the kill cited coal mine discharges of briny water as potential contributing causes but said the algae can also be spread by migratory birds, fishermen, and industrial equipment. PA regulators say they still have not ruled out fracking fluid as a potential contributor but mine drainage, agriculture runoff, and other industrial discharges are also a potential cause.	NO, BUT HANDLING OF FLOWBACK WATER FROM FRACTURING AT ISSUE	UNCLEAR

Well Location	When	Incident and Company Involved	Assessment of Incident	Is Underground Hydraulic Fracturing a Direct Cause?	Are Other Drilling Practices at Issue?
Lower Monongahala River, southwest PA	2008	**COMPANY LINK UNCLEAR.** The U.S. Army Corps of Engineers sounded alarms when the salt level (or "total dissolved solids" level) spiked "dramatically" on the river in October 2008, according to a letter from the Corps to EPA. Although low rain, acid mine drainage, and industrial discharge can also increase salinity of water, it cited "increased gas drilling in the Marcellus Shale" as an aggravating factor. Earlier this year it urged PA to stiffen water treatment standards, saying "conditions are reversing on Pennsylvania's rivers" and it was becoming apparent that the ability of some rivers to receive more salt content was near its limits "and simply cannot sustain" additional levels as a result of gas drilling.	PA environmental regulators cited the Corps' river results and the Dunkard Creek fish kill in April to call for more stringent rules on treatment of discharge water. The new, tougher standards are incentivizing more companies to recycle flowback water rather than treat and dispose of it in PA rivers and streams.	NO, BUT HANDLING OF FLOWBACK WATER FROM FRACTURING AT ISSUE	YES

(Continued)

Well Location	When	Incident and Company Involved	Assessment of Incident	Is Underground Hydraulic Fracturing a Direct Cause?	Are Other Drilling Practices at Issue?
Hopewell Township, southwest PA	2009	**RANGE RESOURCES.** A spill of diluted frack fluid from a Range Resources drilling operation into a small tributary killed small fish, salamanders, and frogs. A relatively small amount of fish were affected, the company said.	PA regulators fined Range $141,175 in May 2010 for the spill. The cause was a broken joint in a transmission line transporting the fluid.	NO, BUT HANDLING OF FLUID ABOVE GROUND AT ISSUE	YES
Dimock, PA, north central PA	2009	**CABOT OIL & GAS.** Cabot Oil & Gas had three spills of fracking water and gel totaling 8,000 gallons within a week. The spills entered a creek and nearby wetland, according to regulatory documents.	PA regulators fined Cabot $56,650 and urged the company to "do a better job in the future of overseeing its contractors." Cabot said the spills were 99.5% water and the material was not hazardous. It said its policy is zero spills.	NO, BUT HANDLING OF FLUID ABOVE GROUND AT ISSUE	NO

15

Notes from the Field

This work would not be balanced if it did not include concerns and complaints from those who claim fracking has affected them adversely. What follows is a small sampling of the overwhelming amount of concern that is being placed locally on the practice of fracking. The following is the synopsis of just one concerned resident, but it speaks volumes as to the depth of passion that is being applied. The following are notes from Kim Feil of Arlington, Texas. The following notes have been reprinted with permission. Content has not been edited or changed in any manner save for typeset for space continuity.

Mon 12/17/2012 7:32 AM

Fr: Kim Feil

We had a drill spill in Lake Arlington, our drinking source, a couple of years ago during a workover. **There should have been setbacks the RRC should have mandated away from our drinking water sources.**

There is also a compressor Station near the lake that an air test uncovered high formaldehyde a couple of years ago that **the RRC**

should have mandated electic compressors station near people and our drinking water source...

http://www.tceq.state.tx.us/assets/public/implementation/barnett_shale/healthEffects/2011.03.16-healthEffectsMemo.pdf

http://www.tceq.state.tx.us/assets/public/implementation/barnett_shale/healthEffects/2011.03.16-healthEffectsMemo.pdf

We also have about 100 drilling laterals under our lake that are at migration risk from seismic events and or cement failures. **The RRC should have never allowed laterals to go UNDER our drinking water source.**

We have had maybe a dozen gas release emission events over the last couple of years that I am aware of. ...**the RRC didn't mandate the TCEQ to test for methane on their sumas so that waste is not occurring in minerals management that the RRC was to protect.**

1. The **RRC should have mandated electric (not diesel) rigs and compressors in ALL urban areas....** http://www.youtube.com/watch?list=FLw_o2P355EYmh7onTqiwcEA&feature=player_embedded&v=js5vVSpWwEw
2. **The RRC should have mandated that a RRC be present during every cement pour or at least video tape ALL cement casing pours** when it comes back up to the top through the annulus so that there is proof of an even pour and **(PER DALE HENRY) the RRC should ensure all wells have electric bond log tests.**
3. **The RRC should have mandated that all drilling mud farming and brine road spraying should be tested and subjected to open records of water and soil test results...** http://www.youtube.com/watch?v=ZQTtI94GGd8&feature=player_embedded Brine spraying
4. **We needed the RRC to seek or mandate the industry to invent technology to keep the toxic, silica dust on the padsite** - those pathetic pillow case looking socks aren't getting the job done. OSHA and NIOSH are aware of the problem... http://blogs.cdc.gov/niosh-science-blog/2012/05/silica-fracking/ http://www.youtube.com/watch?v=SF7fkgzmgO0&feature=youtu.be

5. **We needed the RRC to mandate ventless, emission free flowback tanks by using pressurized flowback tanks instead of open hatch frack tanks....**
   http://www.youtube.com/watch?v=fdKoeBvUHys&feature=player_embedded
6. Global Warming isn't waiting 2.5 years for the EPA mandated Green Completions equipment.... **the RRC should have mandated no venting or flaring period.**
7. **The RRC should have mandated that the pipeline should be in place FIRST before fracturing so that flowback doesn't sit in the ground for months festering bacterial sulfide growth to sour and damage the well and sicken people.....**
   http://barnettshalehell.wordpress.com/2012/03/05/slammed-with-sudden-nausea-from-flowback-odors-business-womans-blood-pressure-soars-after-vomiting/
8. **The RRC should have done health and environmental impact studies before allowing urban drilling.** The setback away from people should be substantial. Rural method drilling is not acceptable in urban areas. An environmental tester who has a Phd said that the health effects/leukemias are being seen downwind from about 1,800 – 2,500 feet....
   http://barnettshalehell.wordpress.com/2012/09/05/envir-tester-ph-d-master-public-health-on-fracking-emissions-fallout-distance-1800-2500-feet-of-the-downwind-effect/
9. **The RRC should have zero tolerance for underinspected, or faked Waste Disposal Injection Well casing pressure tests.** Don't risk eventual migration of toxic fluids into our drinking supplies.....
   http://www.propublica.org/article/trillion-gallon-loophole-lax-rules-for-drillers-that-inject-pollutants
10. **The RRC should regulate how close old wells are to new wells....**

From: Karen Sanchez <Karen.Sanchez@rrc.state.tx.us>
To: kim feil <kimfeil@sbcglobal.net>

Sent: Mon, October 22, 2012 2:17:16 PM
Subject: RE: open records needed to verify when/if well has been shut in permanently

Ms. Feil:

The Railroad Commission of Texas does not regulate how close fracking can occur near old wells.

Kim Feil

http://barnettshalehell.wordpress.com/

TEX LG. CODE ANN. A§ 253.005: Texas Statutes – Section 253.005: LEASE OF OIL, GAS, OR MINERAL LAND
"(c) A well may not be drilled in the thickly settled part of the municipality.."

Texas Administrative Code, Title 30, Part 1, Chapter 101, Subchapter A,Rule 101.4, Environmental Quality, Nuisance

The following are a short list of incidents and concerns generated by Kim.

- Lorrie Squibb Location: Flower Mound, TX Gas Facility: Intensive gas production Exposure: Air Symptoms: Multiple myeloma, a blood cancer http://www.scribd.com/doc/75887335/CWA-Health-Workers-Letter-to-GovCuomo-Fracking-2011-12- 121 http://www.dentonrc.com/local-news/special-projects/gas-well-drilling-headlines/20110831-breast-cancer-rate-climbs-up.ece
- Megan Collins Location: Dish, TX Gas Facility: Compressor station Exposure: Air Symptoms: Sinus issues, difficulty with balance and standing, headaches, fainting, dystonia and ataxia, nausea http://www.earthworksaction.org/voices/detail/dish_texas http://www.npr.org/templates/story/story.php?storyId=120043996
- Deborah Rogers Location: Forth Worth, TX Gas Facility: Chesapeake drilling operations Exposure: Water,

air- benzene, dichlorodifluoromethane, chloroform, xylenes, toluene, disulfides Symptoms: Nausea from the strong odors, nose bleeds, severe headaches. Symptoms (animal): Asphyxiated goats and chickens http://www.earthworksaction.org/voices/detail/deborah_rogers http://vimeo.com/40547126

- Sandra DenBraber Location: Arlington, TX Gas Facility: Carrizo natural gas operations Exposure: Air – benzene, toluene, ethylbenzene, and xylenes Symptoms: Blood has Ethylbenzene, m,p-Xylene, Hexane, 2-Methylpentane, 3-Methylpentane; migraines http://www.texassharon.com/2010/12/14/is-natural-gas-drilling-and-production-making-you-sick/ http://blog.uta.edu/sustainability/2008/04/02/arlington-woman-claims-drill-rig-causessickness/http://www.fwweekly.com/index.php?option=com_content&view=article&id=2906:breathers- beware&catid=76:metropolis&Itemid=377
- Jean Stephens Location: Arlington, TX Gas Facility: Flow back job Exposure: Air Symptoms: Sick to her stomach, disoriented and difficulty breathing http://barnettshalehell.wordpress.com/2012/03/05/slammed-with-sudden-nausea-from-flowback-odors-business-womans-blood-pressure-soars-after-vomiting/
- Tim and Christine Ruggiero and family Location: Wise County, TX Gas Facility: Gas wells Exposure: Water-methane, as well as sodium and other heavy metals Symptoms: Loss of sensation in his extremities, rashes, nausea and memory loss; asthma http://www.earthworksaction.org/voices/detail/tim_and_christine_ruggiero
- Steve Birchfield Location: Johnson County, TX Gas Facility: Gas production Exposure: Water Symptoms: Sick, weak http://www.rodale.com/natural-gas-drilling?page=0,3
- Lloyd Burgess Location: TX Gas Facility: Compressor station Exposure: Air Symptoms (animal): Horses-sick, death, neurological defect and blind in both eyes http://www.npr.org/templates/story/story.php?storyId=120043996

- Charles Morgan Location: Dish, TX Gas Facility: Compressor station Exposure: Air pollution Symptoms: Constant low frequency roar from the compressors has twice ruptured right ear drum; a neighbor who lives on the other side of the compressor station confirmed he and his infant daughter also suffered ruptured ear drums http://www.npr.org/templates/story/story.php?storyId=120043996
- Sharon Ward **(deceased 2011)** Location: Dish, TX Gas Facility: compressor stations, 130 in the County Exposure: Air pollution Symptoms: Unknown, replacement water or relocated http://www.npr.org/templates/story/story.php?storyId=120043996 * Headaches, dizziness, ringing of the ears, and lack of sleep, are symptoms county wide
- Kenneth Bateman and son Location: Justin, TX Gas Facility: compressor station, high pressure pipeline Exposure: Air pollution; noise Symptoms: Diabetes, other health ailments http://www.elkcapital.net/screamingsilence/voices/bateman.html
- Kim Davis Location: Southlake, TX Gas Facility: Gas production Exposure: Air-twenty-six chemicals, carbon disulfide, benzene and naphthalene, carbonyl sulfide, dimethyl disulfide and Pyridine Symptoms: Unknown, replacement water or relocated http://www.dallasobserver.com/2011-11-24/news/fear-and-fracking-in-southlake/2/
- Warren and Rebekah Sheffield and family Location: Dish, TX Gas Facility: Compressor station Exposure: Air, Symptoms: Sick, multiple chemical sensitivity. Vertigo; children easily winded , vomiting, hives, itchy skin, nosebleeds, seizure http://www.dentonrc.com/local-news/special-projects/gas-well-drilling-headlines/20110328- atmosphere-of-concern.ece
- Chuck and Geri Pegg Location: Dish, TX Gas Facility: Few hundred feet northeast of large compressor Exposure: Air – benzene, carbon disulfide, a neurotoxin. Symptoms: Unknown, replacement water or relocated http://www.dentonrc.com/local-news/special-projects/gas-well-drilling-headlines/20110328-atmosphere-of-concern.ece

- Jim and Judy Caplinger Location: Dish, TX Gas Facility: Compressor station Exposure: Air – noxious fumes Symptoms: Unknown, replacement water or relocatedhttp://www.dentonrc.com/local-news/special-projects/gas-well-drilling-headlines/20110328-atmosphere-of-concern.ece
- Steve Lipsky Location: Parker County, TX Gas Facility: Wells, fracking under home Exposure: Water – methane Symptoms: Fatigued and nauseated http://www.dallasobserver.com/2012-04-26/news/fire-in-the-hole/
- Cecil and Tyler Williams Location: Caddo Parrish, LA Gas Facility: Chesapeake spill Exposure: Oil; water – elevated chlorides, oil and grease, organic compounds in soil and water tests Symptoms (animal): **17 dead cows** http://shaleshock.org/2009/08/hours-passed-before-cow-deaths-reported/
- Kelly Gant Location: Bartonville, TX Gas Facility: Compressor station and a gas well near house Exposure: Air Symptoms: Severe asthma attacks, dizzy spells and headaches https://dontfractureillinois.org/Other_communities.html
- Jane Lynn Location: Arlington, TX Gas Facility: Chesapeake natural gas well Exposure: Air Symptoms: Uncontrolled coughing, heart palpitations, burning noses & eyes http://durangotexas.blogspot.com/2011/04/residents-gassed-during-last-nights.html
  http://www.earthworksaction.org/media/detail/arlington_residents_challenge_chesapeake_energy_to_prove_they_are_releasing
- Brian Beadle Location: Hill County, TX Gas Facility: Williams' gas wells Exposure: Water- sulfates, toluene Symptoms (animal): Goats, llama- swelling, death http://archive.fwweekly.com/content.asp?article=6885
- John Sayers Location: Hill County, TX Gas Facility: Williams' gas wells Exposure: Water- sulfates, toluene, hydrocarbons Symptoms: Unknown, replacement water or relocated
  http://archive.fwweekly.com/content.asp?article=6885

- Stevan and Charlotte Harris Location: Hill County, TX Gas Facility: Williams' gas wells Exposure: Water- sulfates, toluene, hydrocarbons Symptoms: Unknown, replacement water or relocated http://archive.fwweekly.com/content.asp?article=6885
- Doug and Diana Harris Location: Denton County, TX Gas Facility: Devon Energy gas production Exposure: Water – high levels of metals: aluminum, arsenic, barium, beryllium, calcium, chromium, cobalt, copper, iron, lead, lithium, magnesium, manganese, nickel, potassium, sodium, strontium, titanium, vanadium Symptoms: Emotional harm and mental anguish, loss of peace of mind, anxiety, and their bodies have been physically injured https://docs.google.com/viewer?a=v&q=cache:A5v6WzCWFLsJ:press.wturley.com/Harris%2520v.%2520Devon%2520Energy.pdf+&hl=en&gl=us&pid=bl&srcid=ADGEESicKmPV6cVx_ihmctJfgSvgEmM7Y-fJSLgm48mUKn_bYfUIunRLuEgIDW5rNjtGkx4JC-zXGZnjnzPaR6NBzbSkAiA1ew8QolwS0EywILW9jHoyJ1rrg06yLEnp164ycN565hCk&sig=AHIEtbTBhDjnA5UCHGqEJrGiBun3LTU uBA
- Grace Mitchell Location: Johnson County, TX Gas Facility: Chesapeake Energy gas production Exposure: Water Symptoms: Unknown, replacement water or relocated http://dfw.cbslocal.com/2010/12/15/north-texas-residents-lawsuits-claim-gas-drilling-contaminated- water/130.
- Jim and Linda Scoma Location: Johnson County, TX Gas Facility: Chesapeake Energy gas production Exposure: Water Symptoms: Unknown, replacement water or relocated http://dfw.cbslocal.com/2010/12/15/north-texas-residents-lawsuits-claim-gas-drilling-contaminated- water/
- John and Jayme Sizelove Location: Denton County, TX Gas Facility: William's drilling operations and compressor stations Exposure: Air Symptoms: headaches, respiratory problems, and other symptoms http://earthjustice.org/features/campaigns/fracking-damage-cases-and-industry-secrecy

- Bob and Lisa Parr, daughter Emma Location: Wise County, TX Gas Facility: 21 gas wells Exposure: Air
- Symptoms: Lisa – rashes, breathing difficulties, nausea and headaches, balance and other neurological problems; Bob – nosebleeds, balance and other neurological problems; Emma – asthma, rashes and nausea http://responsiblegold.com/TXOGAP-CaseStudy-Parr.cfm
- Brian Norberg **(deceased)** Location: Tunnel City, TX Gas Facility: Procore frack sand mining Symptoms: "He felt completely defeated that he could not protect the community from them moving in and destroying our lives," recalls Lisa. He died of a heart attack less than a day later at the age of 52. The family is convinced his death was a result of the stress caused by the conflict. That stress is certainly all too real." http://www.motherjones.com/environment/2012/05/how-rural-america-got-fracked?page=2
- Sinikka Dickerson, Corporal Dale Horton of Arlington Police Dept., Bonnie Billado & daughter Location: Arlington, TX Gas Facility: Unknown which drill-site, winds carried from many drill sites in that area Exposure: Air
- Symptoms: Disoriented, dizzy, & nauseated within a few seconds of odor, lingering headache hours afterwards http://arlingtontx.granicus.com/ViewPublisher.php?view_id=2 Select Dec 6 2011 evening meeting, go to 1:09:17 thru 1:17:45 of City Council meeting for relevant testimonies
- David Hudson Location: Panola County, Texas Gas Facility: Injection wells Exposure: Water - arsenic, cadmium, lead, benzene and other substances Symptoms: Unknown, replacement water or relocated http://www.propublica.org/article/polluted-water-fuels-a-battle-for-answers
- Frank and Earnestene Roberson Location: Panola County, Texas Gas Facility: Injection wells Exposure: Water - arsenic, cadmium, lead, benzene and other substances Symptoms: Unknown, replacement water or relocated http://www.nytimes.com/2006/07/09/us/09deberry.html?pagewanted=print

- Maggie Golden (deceased) Location: Panola County, Texas Gas Facility: Injection wells Exposure: Water - arsenic, cadmium, lead, benzene and other substances Symptoms: Unknown, replacement water or relocated http://www.nytimes.com/2006/07/09/us/09deberry.html?pagewanted=print
- J.D. Johnson Location: Tarrant County. TX Gas Facility: Fracked well Exposure: Water Symptoms: Unknown, replacement water or relocated http://www.mcclatchydc.com/2010/09/07/100188/in-texas-search-for-truth-about.html
- Amber and Damon Smith Location: Denton County, TX Gas Facility: Devon Energy fracked gas well Exposure: Water - arsenic, chromium, butanone, acetone, carbon disulfide, and strontium Symptoms: Unknown, replacement water or relocated http://www.dentonrc.com/local-news/special-projects/gas-well-drilling-headlines/20110329-just- below-the-surface.ece
- Carol Grosser Location: Edwards County, TX Gas Facility: Gas well Exposure: Water Symptoms: Unknown, replacement water or relocated http://www.shalegas.energy.gov/resources/060211_earthworks_natural_gas_flowback.pdf
- Tom and Barbara Vastine Location: Parker County, TX Gas Facility: Gas well Exposure: Water
- Symptoms: Unknown, replacement water or relocated http://www.dentonrc.com/local-news/special-projects/gas-well-drilling-headlines/20110329-just-below-the-surface.ece
- Catherine and Brett Bledsoe Location: Wise County, TX Gas Facility: Aruba Petroleum, fracked wells Exposure: Water - benzene and very high levels of MTBE Symptoms: Water stung their eyes and had odor, animals refused to drink http://www.shalegas.energy.gov/resources/060211_earthworks_natural_gas_flowback.pdf
- Myra, Mike and Cameron Cerney Location: Karnes County, TX Gas Facility: Marathon gas production Exposure: Air
- Symptoms: Frequent headaches, nosebleeds, rashes http://www.texassharon.com/2012/07/23/

letter-from-the-eagle-ford-shale-tells-of-rashes-nose-bleeds- and-misery/ *Many neighbors suddenly have rashes

- Julia Trigg Crawford Location: TX Gas Facility: Pipeline Exposure: N/A Symptoms: Land condemned by eminent domain http://www.huffingtonpost.com/2012/08/23/keystone-pipeline-texas- crawford_n_1824939.html?utm_hp_ref=green
- Michael and Susan Knoll Location: Wise County, TX Gas Facility: Gas wells Exposure: Water, spills on land Symptoms: Coughing Symptoms (animal): Dog – rare form of cancer http://fuelfix.com/blog/2011/05/23/fracking-feud-pits-prosperous-n-texans-against-energy-industry/
- Kim Triolo Feil Location: Arlington TX Gas Facility: Carrizo UT Arlington pad-site & General Motors Natural Gas pad-sites Exposure: Air Symptoms: Increased breathing difficulty, uncontrolled eyebrow twitching/jar clenching, gastro issues, and joint pain/muscle weakness - joint pain/muscle weakness issues relieved with six month nutritionist/detox treatment Symptoms (animal): Parrot developed cough/asthma http://barnettshalehell.wordpress.com/2011/07/24/barnett-shale-arlington-tx-health-effects-diary- natural-gas-urban-drilling-emissions-and-the-canary-effect/ http://www.youtube.com/watch?v=PVxPnJUCDb0
- Josh Beckett Location: McAllen, Texas Gas Facility: Eagle Ford Midstream pipeline Exposure: Loss of habitat due to a failure to adopt an environmentally responsible route that will not cause a prohibited 'take' of the endangered ocelot Symptoms (animal): Endangered ocelot – habitat loss http://www.washingtonpost.com/sports/pitcher-josh-becketts-company-sues-pipeline-builder-over- texas-habitat-for-endangered-ocelot/2012/10/17/1e17836a-1863-11e2-a346-f24efc680b8d_story.html
- Hector Rodriguez Location: Gardendale, Texas Gas facility: Drilling pad Exposure: N/A Symptoms: Disrupted life, ruined plans for building home, property threatened http://www.youtube.com/watch?v=ioMyrTQrkTg

- Ron and Betty Haley Location: Gardendale, Texas Gas Facility: Drilling pad Exposure: N/A Symptoms: Ruined retirement plans for building home, stress http://www.youtube.com/watch?v=EXYZnI3xkcY&feature=relmfu
- Anonymous Location: Gardendale, Texas Gas Facility: Drilling operation Exposure: Land – buried waste pit benzene 40 times EPA standards Symptoms: Evacuated from fumes, water – toxic time bomb http://www.youtube.com/watch?v=D1EFpXrN1gI&feature=relmfu
- Anonymous Location: Gardendale, Texas Gas Facility: Drilling operation Exposure: Land; air- noise Symptoms: Sleep and health affected, dream destroyed http://www.youtube.com/watch?v=iFJvAGgWc4U&feature=relmfu
- Dr. Lawrence Voesack Location: Gardendale, Texas Gas Facility: Drilling operations Exposure: Silica dust, H2S fumes Symptoms: Property value destroyed, quality of life diminished http://www.youtube.com/watch?v=ZtbyudumSRU
- William Justiss Location: Lamar County, Texas Gas Facility: Compressor station Exposure: Air - noxious fumes; noise Symptoms: Nausea, vomiting; house shaking from noise www.supreme.courts.state.tx.us/ebriefs/10/10045106.pdf
- Tom and Judy Alspaugh Location: Lamar County, Texas Gas Facility: Compressor station Exposure: Air - yellowish emission, noise Symptoms: Unable to work outside, house shakes, noses burn www.supreme.courts.state.tx.us/ebriefs/10/10045106.pdf
- Joe Donald Mashburn and Judy Mashburn Location: Lamar County, Texas Gas Facility: Compressor station Exposure: Air - Rotten egg smell, noise Symptoms: Exhaust burns the nose, constantly troubled by noise or smell www.supreme.courts.state.tx.us/ebriefs/10/10045106.pdf
- Joe Denton Mashburn and Christine Mashburn Location: Lamar County, Texas Gas Facility: Compressor station Exposure: Air fumes and noise Symptoms: Because of the fumes, they cannot do

anything outside on their property www.supreme.courts.state.tx.us/ebriefs/10/10045106.pdf

- Joseph Justiss Location: Lamar County, Texas Gas Facility: Compressor station Exposure: Air - fumes and noise Symptoms: Cattle will not graze near pump station because of the vibration or smell on the grass www.supreme.courts.state.tx.us/ebriefs/10/10045106.pdf
- Leonard and Marie Popham Location: Live Oak County, TX Gas Facility: Pioneer Natural Resources gas wells Exposure: Land Symptoms: Stripped of ability to make a living on his own property, property rendered useless to him http://www.texassharon.com/2012/11/11/vietnam-veteran-victim-of-pioneer-natural-resources-strong- arm-tactics-in-eagle-ford-shale/
- Toby Frederick Location: DeWitt County, TX Gas Facility: Range Resources, compressor station and wells Exposure: Water – contaminated with benzene compounds Symptoms: Water discolored with a noxious odor, smell caused dizziness; forced to drill a new, shallower water well http://www.texassharon.com/2011/06/13/eagle-ford-shale-water-contamination-show-and-tell/ http://sacurrent.com/texas-fracking-critics-tour-the-eagle-ford-as-complaints-of-contamination-surface- 1.1165133?pgno=2
- Tim and Shannon Pennington Location: Everman Texas emailed Kim Feil on 12/16/2012 "...we have lost our ground water, our roads are destroyed and there are now earthquakes. ....my husband and i have noticed a severe change in our health. all the symptoms that have been reported by other victims of the natural gas drilling process. ...we are dying literally..."
- Jake and Mike Wilfong, Paleo Pinto, Texas 17 cattle dead after rainstorm flooded drinking areas with mining effluents. http://mineralwellsindex.com/topstory/x439048960/Dead-cattle-spur-lawsuit, http://barnettshalehell.wordpress.com/2012/01/18/emails-on-another-cattle-kill-story-near-mineral-wells-tx-june-2010/
- Jan from Arlington, TX says she smells the gas wells and has health issues, is being tested for BTEX,

formaldehyde & carbon disulfide poisoning- lives on Perkins near Lake Arlington's gas wells and compressor station. http://www.youtube.com/watch?v=8_XIe8Bo72k

- Ethan Tempra, Arlington, TX needed bone marrow transplant - Kim Feil investigated and found he lives on Tiffany Lane near the drill sites and compressor station near Lake Arlington, and goes to school at Grace Prep which is also near a drill site at I20 & Park Springs Blvd. http://www.myfoxdfw.com/story/19044568/boys-life-depends-on-bone-marrow-donor
- Aunt of Grogan family in Arlington, TX living near Martin High (where other suspect cancer clusters have been claimed -see Jan #644 case video) thinks the water maybe the cause of 3 of 4 family members having cancer including the dog. Her brother is a geologist with oil and gas industry with a kidney tumor, her sister-in-law has breast cancer, their child has Ewings Sarcoma. Kim Feil suspects the father is bringing home contaminates on his clothing.
- Exxon (XTO) methanol additive handler for biocide treatments, Arlington, TX. His wife mentioned methanol poisoning, she wife complained of foul odor on his work clothes, their 16-year-old daughter has a benign brain tumor and terrible headaches.
- Arlington, TX video of testimony from Tammie Carson regarding nose bleeds of grandbaby, respiratory issues of daughter and her daughter's husband. The anonymous lady complained her 16-year-old son who lives near drilling in Arlington was having seizures from brain lesions. Tammie complained to me "....(left the window open yesterday, Dec 7 2012) inside my house smelled like paint thinner or lacquer, ...had not been here all day ....my back door faces the line of drilling sites on hwy 287 between Little rd and Sublett. " The date of the complaint matched her being downwind from the gaswells with southwest winds that day. http://www.youtube.com/watch?v=qBrbVusbvIM
- Ft Worth Mary Kelleher, farm animals drowned during rainstorm when drillers changed grading near her

property when pipeline was installed. http://www.fwweekly.com/2012/07/25/dammed/

- Dani Hood, Godley, TX - claims well water, earthquakes and poor health affected by drilling in email to Kim Feil on Facebook.
- Dr Hensley, determatologist – Arlington, TX told Kim Feil and witnessed by son Graham that he saw a lot of rashes from industry men working with silica dust.
- Natalie Genco, Colleyville, TX - Nov 2011 water well went bad just after drilling nearby. http://www.youtube.com/watch?v=cfl-0yL4oCA
- Graham Feil age 13, Elm St, Arlington, TX: eyes dilated, lethargic for ten days just after playing outside on a snow "white out" event in Feb 2010. Mom, Kim Feil, suspects cooling inversion pushed pollution down to ground - had headaches, nausea, dizzy, rapid breathing, brain felt scrambled, exhausted yet anxious/hyper. His doctor said he exhibited symptoms of exposure yet no diagnosis was made. Lives downwind to 22 gas wells and two lift compressors at UT Arlington Carrizo gaspad site. http://www.youtube.com/watch?v=7p-G36wAONc&feature=player_embedded, http://barnettshalehell.wordpress.com/2011/07/24/barnett-shale-arlington-tx-health-effects-diary-natural-gas-urban-drilling-emissions-and-the-canary-effect/
- Bonnie, Arlington, TX UTA Community Gardens blames two extreme nosebleeds in Jan & Feb 2012 on nearby drilling http://www.youtube.com/watch?v=ztBRDsojySE&feature=player_embedded
- Pantego, TX, five wells fracked near Bailey Junior High, Kim Feil reports during that time Graham Feil, Bailey student, diagnosed with asthma, Mrs. Stuteville, a Bailey teacher, deceased from aneurysm just after the fracking and that by the summer, Bailey music teacher, Mrs Lynn Wilson, continues to battle 4th stage bone cancer (Wilson also lives downwind to the Arlington Compressor Station on Denham Dr). http://www.youtube.com/watch?v=Ht1RU_7r4KI&feature=player_embedded

- Mary Owen, East St, Arlington, TX near Chesapeake, Truman drill site hospitalized for worsened fibromyalgia summer 2012 during fracking of 3 wells.
- Walker age 14, Arlington, TX April 2011 following the Fulson gas emission event (who also lives about 1,200 feet downwind from the Ragland Site), was hospitalized hours after the gas leak with a 105 fever and swollen neck/enlarged lymph nodes. No diagnosis was made.
- Regina & granddaughter, Spitfire Dr, Arlington, TX downwind from Arlington Compressor Station told me she & her granddaughter have respiratory issues ever since drilling came in. Three ER visits in November, 2010 and grandbaby's nosebleeds are now 3 times a day. She told me sometimes she smelled fumes in the girl's bedroom. There also is a drill site to the north of them too.

15.1 Going Forward

In one respect, the concerns raised could be easily debated and put to rest. It has been suggested that if the well production companies stay within the letter of the law and EPA requirements, then many, if not all, problems can be solved, and future incidents of illness and discomfort shall diminish. As in so many instances, the public perception, the media's spin, and the industries' explanation rarely converge. One would never have imagined that a fourth-grade math diagram would come in handy, but it is very apparent that this scenario plays out like a highly defined Venn Diagram with each faction holding on dearly to their circle of 'truth.'

The temptation to take every single concern and debate to resolution is strong, yet it would be a fool's errand. Concerns mentioned are very real, if in no other eyes than the conveyor. What has to be done is to make a strong and convincing argument to continue the pressure on increased safe exposure and reduced environmental impact.

One hundred fifty years ago when crude oil was first being extracted, the damage done to the environment was nothing short of a nightmare. In some areas, little has changed, but in many instances, companies take extraordinary precautions. Many years ago, the ramifications of pollution were not understood. Today, much work is underway to address what is being understood as environmental concerns. The work that goes into preparing a fracking well site today in assuring that the chemicals used are innocuous while maintaining the physical integrity of the surrounding land is taken into consideration primarily for legal reasons as well as business concerns. While many opponents believe that exploration, drilling, and product producing companies will stop at nothing to turn a profit, this is simply not the case. If profit was the only motivation, more American companies would be in business as well as in court. Companies for the most part want to do the right thing. If environmental laws are in place, then work shall be structured accordingly. The challenge is to enact laws that make sense according to empirical evidence. It is also vital that those using the process and chemicals as well as the public understand the technology. As with all moral and ethical debates, evidence, albeit empirical, is all that is available. One cannot make decisions based on hunch or conjecture. One can only apply prudential certitude and act accordingly. Where industry in general fails is rushing forward and only applying the letter of the law and not working to push for higher requirements. Legislation is often left to the most verbal and most passionate. It is fair to say that when emotion runs high, logic wanes. If we are to utilize the gifts that the earth has bestowed upon us, then it is more than fair to assume a protective role going forward and to explore proper protocols to ensure that the environment and comfort of life remain as balanced as it was found. All can learn from the Boy Scout ethic of *Leave No Trace.*

The authors would strongly suggest that if the reader has any concerns, ideas, or questions to direct them to local government agencies. The government is made up of people that may live in one's town or city. They have the same concerns as those in opposition yet they may not be aware of certain incidents. Please bring concerns to them so that the government officials will be made aware.

One last request, anyone who engages in an argument that ends with the statement "But what about the children…" shall be forever be thought of as weak and manipulative. Let us base our arguments on facts that can be substantiated. Our concerns will be better received, and just maybe the point will be proven and acted upon.

Appendix A

Chemicals Used in Fracking

The following is a list of the common chemicals used at a fracking site. One of the problems associated with identifying chemicals is that some chemicals have multiple names. For example, Ethylene Glycol (Antifreeze) is also known by the names Ethylene alcohol; Glycol; Glycol alcohol; Lutrol 9; Macrogol 400 BPC; Monoethylene glycol; Ramp; Tescol; 1,2-Dihydroxyethane; 2-Hydroxyethanol; HOCH2CH2OH; Dihydroxyethane; Ethanediol; Ethylene gycol; Glygen; Athylenglykol; Ethane-1,2-diol; Fridex; M.e.g.; 1,2-Ethandiol; Ucar 17; Dowtherm SR 1; Norkool; Zerex; Aliphatic diol; Ilexan E; Ethane-1,2-diol 1,2-Ethanedio.

This multiplicity of names can make a search for chemicals somewhat difficult and frustrating. However, if you search for a chemical by the CAS number, it will return the correct chemical even if the name on the fracturing record does not match. For example, if the fracturing record listed the chemical Hydrogen chloride and one searched for it by name using a chemical search site, you may not get a result. But if a search for CAS # 007647-01-0, it might return Hydrochloric acid, which is another name of Hydrogen chloride. Therefore, by using the CAS number, one can avoid the issue of multiple names for the same chemical.

Chemical Name	CAS	Chemical Purpose	Product Function
Hydrochloric Acid	007647-01-0	Helps dissolve minerals and initiate cracks in the rock	Acid
Glutaraldehyde	000111-30-8	Eliminates bacteria in the water that produces corrosive by-products	Biocide
Quaternary Ammonium Chloride	012125-02-9	Eliminates bacteria in the water that produces corrosive by-products	Biocide
Quaternary Ammonium Chloride	061789-71-1	Eliminates bacteria in the water that produces corrosive by-products	Biocide
Tetrakis Hydroxymethyl-Phosphonium Sulfate	055566-30-8	Eliminates bacteria in the water that produces corrosive by-products	Biocide
Ammonium Persulfate	007727-54-0	Allows a delayed break down of the gel	Breaker
Sodium Chloride	007647-14-5	Product Stabilizer	Breaker
Magnesium Peroxide	014452-57-4	Allows a delayed break down the gel	Breaker
Magnesium Oxide	001309-48-4	Allows a delayed break down the gel	Breaker
Calcium Chloride	010043-52-4	Product Stabilizer	Breaker
Choline Chloride	000067-48-1	Prevents clays from swelling or shifting	Clay Stabilizer
Tetramethyl ammonium chloride	000075-57-0	Prevents clays from swelling or shifting	Clay Stabilizer

Chemical Name	CAS	Chemical Purpose	Product Function
Sodium Chloride	007647-14-5	Prevents clays from swelling or shifting	Clay Stabilizer
Isopropanol	000067-63-0	Product stabilizer and / or winterizing agent	Corrosion Inhibitor
Methanol	000067-56-1	Product stabilizer and / or winterizing agent	Corrosion Inhibitor
Formic Acid	000064-18-6	Prevents the corrosion of the pipe	Corrosion Inhibitor
Acetaldehyde	000075-07-0	Prevents the corrosion of the pipe	Corrosion Inhibitor
Petroleum Distillate	064741-85-1	Carrier fluid for borate or zirconate crosslinker	Crosslinker
Hydrotreated Light Petroleum Distillate	064742-47-8	Carrier fluid for borate or zirconate crosslinker	Crosslinker
Potassium Metaborate	013709-94-9	Maintains fluid viscosity as temperature increases	Crosslinker
Triethanolamine Zirconate	101033-44-7	Maintains fluid viscosity as temperature increases	Crosslinker

(*Continued*)

Chemical Name	CAS	Chemical Purpose	Product Function
Sodium Tetraborate	001303-96-4	Maintains fluid viscosity as temperature increases	Crosslinker
Boric Acid	001333-73-9	Maintains fluid viscosity as temperature increases	Crosslinker
Zirconium Complex	113184-20-6	Maintains fluid viscosity as temperature increases	Crosslinker
Borate Salts	N/A	Maintains fluid viscosity as temperature increases	Crosslinker
Ethylene Glycol	000107-21-1	Product stabilizer and / or winterizing agent.	Crosslinker
Methanol	000067-56-1	Product stabilizer and / or winterizing agent.	Crosslinker
Polyacrylamide	009003-05-8	"Slicks" the water to minimize friction	Friction Reducer
Petroleum Distillate	064741-85-1	Carrier fluid for polyacrylamide friction reducer	Friction Reducer
Hydrotreated Light Petroleum Distillate	064742-47-8	Carrier fluid for polyacrylamide friction reducer	Friction Reducer
Methanol	000067-56-1	Product stabilizer and / or winterizing agent.	Friction Reducer

Chemical Name	CAS	Chemical Purpose	Product Function
Ethylene Glycol	000107-21-1	Product stabilizer and / or winterizing agent.	Friction Reducer
Guar Gum	009000-30-0	Thickens the water in order to suspend the sand	Gelling Agent
Petroleum Distillate	064741-85-1	Carrier fluid for guar gum in liquid gels	Gelling Agent
Hydrotreated Light Petroleum Distillate	064742-47-8	Carrier fluid for guar gum in liquid gels	Gelling Agent
Methanol	000067-56-1	Product stabilizer and / or winterizing agent.	Gelling Agent
Polysaccharide Blend	068130-15-4	Thickens the water in order to suspend the sand	Gelling Agent
Ethylene Glycol	000107-21-1	Product stabilizer and / or winterizing agent.	Gelling Agent
Citric Acid	000077-92-9	Prevents precipitation of metal oxides	Iron Control
Acetic Acid	000064-19-7	Prevents precipitation of metal oxides	Iron Control
Thioglycolic Acid	000068-11-1	Prevents precipitation of metal oxides	Iron Control
Sodium Erythorbate	006381-77-7	Prevents precipitation of metal oxides	Iron Control

(*Continued*)

Chemical Name	CAS	Chemical Purpose	Product Function
Lauryl Sulfate	000151-21-3	Used to prevent the formation of emulsions in the fracture fluid	Non-Emulsifier
Isopropanol	000067-63-0	Product stabilizer and / or winterizing agent.	Non-Emulsifier
Ethylene Glycol	000107-21-1	Product stabilizer and / or winterizing agent.	Non-Emulsifier
Sodium Hydroxide	001310-73-2	Adjusts the pH of fluid to maintains the effectiveness of other components, such as crosslinkers	pH Adjusting Agent
Potassium Hydroxide	001310-58-3	Adjusts the pH of fluid to maintains the effectiveness of other components, such as crosslinkers	pH Adjusting Agent
Acetic Acid	000064-19-7	Adjusts the pH of fluid to maintains the effectiveness of other components, such as crosslinkers	pH Adjusting Agent
Sodium Carbonate	000497-19-8	Adjusts the pH of fluid to maintains the effectiveness of other components, such as crosslinkers	pH Adjusting Agent

Chemical Name	CAS	Chemical Purpose	Product Function
Potassium Carbonate	000584-08-7	Adjusts the pH of fluid to maintains the effectiveness of other components, such as crosslinkers	pH Adjusting Agent
Copolymer of Acrylamide and Sodium Acrylate	025987-30-8	Prevents scale deposits in the pipe	Scale Inhibitor
Sodium Polycarboxylate	N/A	Prevents scale deposits in the pipe	Scale Inhibitor
Phosphonic Acid Salt	N/A	Prevents scale deposits in the pipe	Scale Inhibitor
Lauryl Sulfate	000151-21-3	Used to increase the viscosity of the fracture fluid	Surfactant
Ethanol	000064-17-5	Product stabilizer and / or winterizing agent.	Surfactant
Naphthalene	000091-20-3	Carrier fluid for the active surfactant ingredients	Surfactant
Methanol	000067-56-1	Product stabilizer and / or winterizing agent.	Surfactant
Isopropyl Alcohol	000067-63-0	Product stabilizer and / or winterizing agent.	Surfactant
2-Butoxyethanol	000111-76-2	Product stabilizer	Surfactant

The following are the Safety Data Sheets for the typical chemicals found at a fracing site.

Acetaldehyde	
• Ethyl aldehyde • Acetic aldehyde • Ethanal • Acetylhydride • Aldehyde • AAD	
Formula	CH_3CHO
Structure	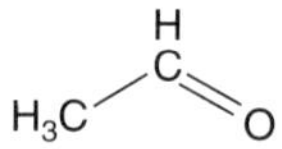
Description	Highly volatile, colorless liquid with a penetrating, pungent, suffocating odor that is somewhat fruity and quite pleasant in low concentrations.
Uses	Acetaldehyde is used mainly as a chemical intermediate in the production of acetic acid. It is also used to produce many other chemicals. It is used in the manufacture of paraldehyde, metaldehyde, other polymers, plastics, synthetic rubber and resins, cosmetics, perfumes, pesticides and pharmaceuticals; in the silvering of mirrors; in leather tanning; in the hardening of gelatin fibres; as a denaturant for alcohols; in fuel compositions; in glue and casein products; as a preservative for fish and fruit; as a synthetic flavouring agent; as a food additive; in the paper industry; and as a laboratory chemical.

Registry Numbers and Inventories	
CAS	75-07-0
NIH PubChem CID	177
EC (EINECS/ELINCS)	200-836-8
EC Index Number	605-003-00-6
EC Class	F+; R12, Carc. Cat. 3; R40, Xi; R36/37
RTECS	AB1925000
RTECS class	Tumorigen; Mutagen; Reproductive Effector; Human Data; Primary Irritant
UN (DOT)	1089
Merck	12,37
Beilstein/Gmelin	505984
Beilstein Reference	4-01-00-03094
RCRA	U001
EPA OPP	202300
FEMA	2003
Swiss Giftliste 1	G-1024
Canada DSL/NDSL	DSL
US TSCA	Listed
Austrailia AICS	Listed
New Zealand	Listed
Japan ENCS (MITI)	Listed
Korea ECL	Listed
Philippiens PICCS	Listed

Properties	
Formula	C_2H_4O
Formula mass	44.05
Melting point, °C	-123
Boiling point, °C	20.1
Vapor pressure, mm_{Hg}	965 (25 C)
Vapor density (air=1)	1.52
Saturation Concentration	Extremely high; gas at room temperature.
Odor threshold	Recognition 0.21 ppm
Critical temperature	193
Critical pressure	54.8
Density	0.780 g/cm^3 (20 C)
Solubility in water	Miscible
Viscosity	0.2456 cp at 15C
Surface tension	21.2 g/s2 at 20 C
Refractive index	1.3316 (20 C)
Partition coefficient, pK_{ow}	-0.16
Thermal expansion	0.0021/K at 20 C
Heat of vaporization	25.7 kJ/mol
Heat of combustion	-1169 kJ/mol

Hazards and Protection	
Storage	Keep away from heat, sparks, and flame. Keep away from sources of ignition. Keep from freezing. Store in a tightly closed container. Keep from contact with oxidizing materials. Keep away from strong acids. Refrigerator/flammables. Keep away from reducing agents. Do not expose to air. After opening, purge container with nitrogen before reclosing. Periodically test for peroxide formation on long-term storage. Addition of water or appropriate reducing materials will lessen peroxide formation. Store under an inert atmosphere.
WHMIS	B2 D2A
Handling	Use only in a well ventilated area. Ground and bond containers when transferring material. Do not breathe dust, vapor, mist, or gas. Do not get in eyes, on skin, or on clothing. Empty containers retain product residue, (liquid and/or vapor), and can be dangerous. Keep container tightly closed. Avoid contact with heat, sparks and flame. Do not ingest or inhale. Handle under an inert atmosphere. Store protected from air. This product may be under pressure; cool before opening. If peroxide formation is suspected, do not open or move container. Do not pressurize, cut, weld, braze, solder, drill, grind, or expose empty containers to heat, sparks or open flames.
Protection	Eyes: Wear appropriate protective eyeglasses or chemical safety goggles as described by OSHA's eye and face protection regulations in 29 CFR 1910.133 or European Standard EN166. Skin: Wear appropriate protective gloves to prevent skin exposure. Clothing: Wear appropriate protective clothing to minimize contact with skin.

(*Continued*)

Hazards and Protection	
Respirators	Follow the OSHA respirator regulations found in 29CFR 1910.134 or European Standard EN 149. Always use a NIOSH or European Standard EN 149 approved respirator when necessary.
Small spills/leaks	Absorb spill with inert material, (e.g., dry sand or earth), then place into a chemical waste container. Use water spray to dilute spill to a non-flammable mixture. Avoid runoff into storm sewers and ditches which lead to waterways. Use water spray to disperse the gas/vapor. Remove all sources of ignition. Use a spark-proof tool. Place under an inert atmosphere.
Disposal code	7
Stability	Unstable in air. May undergo autopolymerization. Forms explosive peroxides on prolonged storage and exposure to air.
Incompatibilities	Air, acid anhydrides, alcohols, ketones, phenols, ammonia, hydrogen cyanide, hydrogen sulfide, halogens, phosphorus, isocyanates, stong alkalies, amines, cobalt chloride, mercury (II) chlorate, mercury (II) perchlorate, trace metals, acids, acetic acid, sulfuric acid, strong oxidizing agents, strong reducing agents, strong bases, strong acids.
Decomposition	Carbon monoxide, carbon dioxide, methane.

Fire	
Flash Point,°C	–38
Autoignition, °C	140
Upper exp. limit, %	57
Lower exp. limit, %	4

Fire fighting		Wear a self-contained breathing apparatus in pressure-demand, MSHA/NIOSH (approved or equivalent), and full protective gear. Water runoff can cause environmental damage. Dike and collect water used to fight fire. Use water spray to keep fire-exposed containers cool. Wear appropriate protective clothing to prevent contact with skin and eyes. Wear a self-contained breathing apparatus (SCBA) to prevent contact with thermal decomposition products. Extremely flammable liquid. Vapor may cause flash fire. Forms peroxides of unknown stability. Containers may explode in the heat of a fire. Will be easily ignited by heat, sparks or flame. Extinguishing media: For small fires, use dry chemical, carbon dioxide, water spray or alcohol-resistant foam. For large fires, use water spray, fog, or alcohol-resistant foam. Use water spray to cool fire-exposed containers.
Fire potential		Very flammable, combustion imminent.
Hazards		Vapors are heavier than air and may travel a considerable distance to a source of ignition and flash back.
Combustion products		Produces irritating vapor when heated. Vapors are heavier than air and may travel a considerable distance to a source of ignition and flash back.
NFPA	**Health**	2
	Flammability	4
	Reactivity	2

Health	
Exposure limit(s)	OSHA PEL: TWA 200 ppm (360 mg/m^3) NIOSH REL: Ca See Appendix A See Appendix C (Aldehydes) NIOSH IDLH: Potential occupational carcinogen 2000 ppm

(*Continued*)

Health	
Carcinogen	G-A3, I-2B, N-2, CP65
Poison Class	4
Exposure effects	Prolonged or repeated skin contact may cause dermatitis. Prolonged or repeated eye contact may cause conjunctivitis. May cause cancer according to animal studies. May cause reproductive and fetal effects. Laboratory experiments have resulted in mutagenic effects. Prolonged skin contact may cause erythema and burns.
Ingestion	Harmful if swallowed. May cause gastrointestinal irritation with nausea, vomiting and diarrhea. May cause central nervous system depression.
Inhalation	Causes respiratory tract irritation. May cause narcotic effects in high concentration. Exposure produces central nervous system depression. Vapors may cause dizziness or suffocation. Can produce delayed pulmonary edema. Inhalation of large amounts may cause respiratory stimulation, followed by respiratory depression, convulsions and possible death due to respiratory paralysis. May cause respiratory sensitization.
Skin	May cause skin sensitization, an allergic reaction, which becomes evident upon re-exposure to this material. Causes skin irritation and possible burns.
Eyes	Lachrymator. Causes severe eye irritation and possible burns.

First aid	
Ingestion	Do NOT induce vomiting. If victim is conscious and alert, give 2-4 cupfuls of milk or water. Never give anything by mouth to an unconscious person. Get medical aid immediately.

First aid	
Inhalation	Get medical aid immediately. Remove from exposure to fresh air immediately. If breathing is difficult, give oxygen. DO NOT use mouth-to-mouth respiration. If breathing has ceased apply artificial respiration using oxygen and a suitable mechanical device such as a bag and a mask.
Skin	Get medical aid immediately. Immediately flush skin with plenty of soap and water for at least 15 minutes while removing contaminated clothing and shoes. Discard contaminated clothing in a manner which limits further exposure. Destroy contaminated shoes.
Eyes	Get medical aid immediately. Do NOT allow victim to rub or keep eyes closed. Extensive irrigation is required (at least 30 minutes).

Transportation		
UN number	1089	
Response guide	129	
Hazard class	3.1	
Packing Group	I	
USCG CHRIS Code	AAD	
USCG Compatatibility Group	19 Aldehydes	
HS Code	2912 12 00	
Std. Transport #	4907210	
IMO Pollution Category	C	
IMO Gas Code	B	

OPPT Chemical Fact Sheet EPA 749-F-94-003
CHEMICALS IN THE ENVIRONMENT: ACETALDEHYDE (CAS NO. 75-07-0)
prepared by
OFFICE OF POLLUTION PREVENTION AND TOXICS
U.S. ENVIRONMENTAL PROTECTION AGENCY
August 1994

Chemicals can be released to the environment as a result of their
manufacture, processing, and use. EPA has developed information summaries
on selected chemicals to describe how you might be exposed to these
chemicals, how exposure to them might affect you and the environment, what
happens to them in the environment, who regulates them, and whom to contact
for additional information. EPA is committed to reducing environmental
releases of chemicals through source reduction and other practices that
reduce creation of pollutants.

WHAT IS ACETALDEHYDE, HOW IS IT USED, AND HOW MIGHT I BE EXPOSED?

Acetaldehyde is a colorless, flammable liquid. It occurs naturally in certain foods, such as ripe fruits and coffee, and in cigarette smoke. Green plants produce acetaldehyde as they break down food. It is produced in large amounts (740 million pounds in 1989) by two companies in the United States. US production of acetaldehyde may decline in the near future as Mexico increases its production of the chemical. US demand for acetaldehyde increased slightly from 1987 to 1989, but is likely to remain steady at its current level. The largest users of acetaldehyde are companies that make acetic acid and related chemicals. Companies also use acetaldehyde to make other chemicals such as pyridine, pentaerythritol, and peracetic acid.

Exposure to acetaldehyde can occur in the workplace or in the environment following releases to air, water, land, or groundwater. Exposure can also occur when people eat fruit, drink coffee, or smoke cigarettes. Acetaldehyde enters the body when breathed in with contaminated air or when consumed with contaminated food or water. It does not remain in the body due to its breakdown, mainly to a chemical that is normally present in the human body.

WHAT HAPPENS TO ACETALDEHYDE IN THE ENVIRONMENT?
Acetaldehyde evaporates when exposed to air. It dissolves completely when mixed with water. Most direct releases of acetaldehyde to the environment are to air or to underground sites. It also evaporates from water and soil exposed to air. Once in air, it breaks down to other chemicals. Microorganisms that live in water and in soil can also break down acetaldehyde. Because it is a liquid that does not bind well to soil, acetaldehyde that makes its way into the ground can move through the ground and enter groundwater. Plants and animals are not likely to store acetaldehyde.

HOW DOES ACETALDEHYDE AFFECT HUMAN HEALTH AND THE ENVIRONMENT?

Effects of acetaldehyde on human health and the environment depend on the amount of acetaldehyde present and the length and frequency of exposure. Effects also depend on the health of a person or the condition of the environment when exposure occurs.

Breathing acetaldehyde for short periods of time can irritate the human respiratory system. Acetaldehyde can also adversely affect the cardiovascular system. Contact with acetaldehyde liquid or vapor irritates the skin and the eyes. These effects are not likely to occur at levels of acetaldehyde that are normally found in the environment.

Human health effects associated with breathing or otherwise consuming small amounts of acetaldehyde over long periods of time are not known. Limited evidence from animal studies shows that acetaldehyde can adversely affect the developing fetus. Laboratory studies also show that acetaldehyde can severely damage the respiratory tract and cause cancer in animals exposed repeatedly by inhalation. Repeat exposure to acetaldehyde in air may likewise cause cancer in humans.

Acetaldehyde by itself is not likely to cause environmental harm at levels normally found in the environment. Acetaldehyde can contribute to the formation of photochemical smog when it reacts with other volatile organic carbon substances in air.

CHEMICAL SUMMARY FOR
ACETALDEHYDE EPA 749-F-94-003a

prepared by
OFFICE OF POLLUTION PREVENTION AND TOXICS
U.S. ENVIRONMENTAL PROTECTION AGENCY
August 1994

This summary is based on information retrieved from a systematic search limited to secondary sources (see Appendix A). These sources include online databases, unpublished EPA information, government publications, review documents, and standard reference materials. No attempt has been made to verify information in these databases and secondary sources.

I. CHEMICAL IDENTITY AND PHYSICAL/ CHEMICAL PROPERTIES

The chemical identity and physical/chemical properties of acetaldehyde are summarized in Table 1.

II. PRODUCTION, USE, AND TRENDS

Table 1 Chemical Identity And Chemical/Physical Properties Of Acetaldehyde.

Characteristic/Property	Data	Reference
CAS No.	75-07-0	
Common Synonyms acetic	aldehyde;	
	ethanal;	
ethyl aldehyde	HSDB 1994	

Characteristic/Property	Data	Reference
Molecular Formula	C_2H_4O	
Chemical Structure	CH_3-CHO	
Physical State colorless liquid		Verschueren 1983
Molecular Weight 44.05		Budavari et al. 1989
Melting Point -123.5°C		Budavari et al. 1989
Boiling Point 21°C at 760 mm Hg		Budavari et al. 1989
Water Solubility miscible		Budavari et al. 1989
Density d16/4, 0.788		Budavari et al. 1989
Vapor Density (air = 1)	1.52	ACGIH 1991
KOC		not available
Log KOW	-0.22 (estimated)	CHEMFATE 1994
Vapor Pressure	740 mm Hg @ 20°C	Verschueren 1983
Reactivity	highly reactive; flammable	Budavari et al. 1989;
		Verschueren 1983
Flash Point	-36°F (-38°C)	
	(closed cup)	Budavari et al. 1989
Henry's Law Constant	7.89×10^{-5} atm-m^3/mol	
	at 25°C	CHEMFATE 1994
Fish Bio-concentration		
Factor	<1 (estimated)	CHEMFATE 1994
Odor Threshold	0.21 ppm (in air)	Verschueren 1983
Conversion Factors		1 ppm = 1.83 mg/m^3
	1 mg/m^3= 0.55 ppm	Verschueren 1983

A. Production

There are two producers of acetaldehyde in the United States,Eastman and Hoechst Celanese. In 1989, US production of acetaldehyde was estimated to be 740 million pounds. In the same year, 40 million pounds were exported (Mannsville, 1990).

B. Use

The largest use of acetaldehyde is as a chemical intermediate forthe production of acetic acid. The production of esters, principally ethyl acetate and isobutyl acetate, is the second most significant use. It is also used as an intermediate for a number of other chemicals, including pyridine and pyridine bases, pentaerythritol, peracetic acid, 1,3-butylene glycol, and chloral. Acetaldehyde is also used in silvering mirrors; hardening gelatin fibers; denaturing alcohol; and in the manufacture of disinfectants, dyes, drugs, explosives, flavorings, phenolic and urea resins, rubber accelerators and antioxidants, varnishes, and yeast.

Table 2 shows the estimated 1986 U.S. end-use pattern for acetaldehyde.

C. Trends

Both production and demand increased some what between 1986-1989. Though demand is expected to remain static at about 700 million pounds, US production may decline due to the addition of production capacity in Mexico.

Table 2 Estimated 1986 United States End-Use Pattern of Acetaldehyde.

Use of Acetaldehyde	Percentage of US
(typical Standard Industrial Classification (SIC) Code)	Acetaldehyde Use
Acetic acid (production, SIC 2869)	55%

Use of Acetaldehyde	Percentage of US
Ester production (production, SIC 2861)	15%
Pyridine and pyridine bases (production, SIC 2822)	10%
Pentaerythritol (production, SIC 2869)	10%
Peracetic acid (production, SIC 2869)	5%
1,3-butylene glycol (SIC codes unknown)	2%
Miscellaneous and exports (no applicable SIC Code(s))	3%

Source: Mannsville, 1993.

III. ENVIRONMENTAL FATE

A. Environmental Release

Acetaldehyde is released into air or wastewater from facilities producing or using the chemical (HSDB 1994). Acetaldehyde is also released to the environment from the combustion and photooxidation of hydrocarbons (HSDB 1994). Acetaldehyde is an intermediate product of respiration in higher plants and occurs naturally in many foods, such as ripe fruits that have tart tastes before ripening, and coffee (HSDB 1994). Acetaldehyde is a component of cigarette smoke (HSDB 1994).

In 1992, releases of acetaldehyde to environmental media, as reported to the Toxic Chemical Release Inventory by certain types of U.S. industries, totaled about 8.4 million pounds: 6.42 million pounds to the atmosphere; 1.9 million pounds to underground injection sites; 77,188 pounds to surface water discharges; and 289 pounds to land (TRI92 1994).

Concentrations of acetaldehyde measured in air samples taken from different locations vary, depending on several conditions, including weather. The chemical has been detected in ice fog, rain, cloud mist, and fog (CHEMFATE 1994). Urban measured 5.2 ppb in Baltimore, MD; 35 ppb in Claremont, CA (severe smog); and 170 ppb in another unspecified California city (foggy conditions). Acetaldehyde concentrations reached 590 ppb in clouds over California (site not specified), and levels reached 3.1-13.3 ppb in Grand Canyon, AZ (CHEMFATE 1994). Trace amounts of acetaldehyde have been detected in the drinking water of several U.S. cities (CHEMFATE 1994; HSDB 1994).

B. Transport

The Henry's Law constant for acetaldehyde, 7.89×10^{-5} atm.m^3/ mol at 25°C (CHEMFATE 1994), and its high vapor pressure, 740 mm Hg @ 20°C (Verschueren 1983), indicate that volatilization of the chemical from surface soil or water to the atmosphere will be considerable.

C. Transformation/Persistence

1. Air - In air (at 25°C), acetaldehyde reacts with OH radicals, NO3, singlet oxygen, and NO2 at the following rates: 0.15×10^{-10}, 0.12×10^{-14}, 0.31×10^{-12}, and 0.20×10^{-24} cm^3/molecule-sec, respectively. The estimated half-life for the reaction of acetaldehyde with OH produced by UV light is 6.2 hours; the products of this reaction include peroxyacetylnitrate (PAN), methyl nitrate, methyl nitrite, and nitric acid (CHEMFATE 1994). Acetaldehyde absorbs UV light at wavelengths of 290 to 342 nm (CHEMFATE 1994), indicating some potential for photolysis. The photolytic half-lives for acetaldehyde are about 34 hours in the summer and 296 hours in winter at 55° N latitude. The estimated half-life for the reaction of acetaldehyde with ozone at 25°C, based on the rate constant of 6×10-21 cm^3/molecule-sec, is 1910 days (5.2 years) (CHEMFATE 1994).
2. Soil - Acetaldehyde will volatilize rapidly in near surface and surface soils [v.p. 740 mm Hg at 20°C (Verschueren 1983)], or leach into the ground,

or undergo microbial degradation (HSDB 1994). Acetaldehyde is not expected to adsorb to soils, other than those containing montmorillonite clay (HSDB 1994).

3. Water - If released to water, acetaldehyde will rapidly biodegrade or volatilize (for a typical river, the half-life is 9.3 hours) (HSDB 1994). Laboratory tests demonstrate that acetaldehyde is easily biodegraded (1) by acclimated sludge and sewage with theoretical biological oxygen damand (BOD) values for several studies ranging from 28% in 24 hours to 100% in 5 hours; (2) by anaerobic treatment; and (3) in seawater (25% degradation in 1 hour, compared with no decline in concentration in sterile controls) (HSDB 1994).
4. Biota - The estimated, low KOW and bioconcentration values for acetaldehyde (CHEMFATE 1994) indicate that there is little potential for the bioaccumulation or bioconcentration of acetaldehyde in biota.

IV. HEALTH EFFECTS

Acetaldehyde is the major metabolite of ethanol (HSDB 1994). Many of the adverse effects of ethanol are attributed to acetaldehyde (HSDB 1994). Direct administration of acetaldehyde to rats has established alcohol dependency (Brabec 1981).

A. Pharmacokinetics

1. Absorption - Acetaldehyde is rapidly absorbed by oral and respiratory routes (ACGIH 1991). Estimated half-lives of acetaldehyde in circulating blood have been reported as <15 minutes (ACGIH 1991) and 3.1 minutes (HSDB 1994).
2. Distribution - Experimental studies indicate that, following inhalation or oral exposure, sufficient first-pass metabolism occurs in the liver and respiratory tract to limit acetaldehyde was detectable in the liver, blood, kidney, spleen, heart, and bone of rats exposed to the 20 mM vapor for 1 hour and in maternal and embryonic

tissues following administration (route unspecified) of >5 g/kg ethanol to pregnant mice (HSDB 1994).

3. Metabolism - Acetaldehyde is metabolized (mainly in mammalian liver) to acetic acid by NAD+-dependent aldehyde dehydrogenase. The rate of metabolism to acetic acid varies, but it is generally considered to be rapid (IARC 1985; ACGIH 1991). Saturation kinetics are not apparent even with large doses (ACGIH 1991). Acetic acid enters the metabolic pool of intermediary metabolism and is used in the production of carbon dioxide and water or in cellular synthesis of cholesterol, fatty acids, and other tissue constituents (HSDB 1994). In vitro, acetaldehyde formed adducts with cytosine- and purine-containing nucleotides (IARC 1985).
4. Excretion - Because the major metabolite of acetaldehyde enters into intermediary metabolism, the excretion of the parent compound or its metabolites may be limited. Acetaldehyde has been detected in expired air (usually no more than 5%) but only its metabolites have been detected in the urine (ACGIH 1991). Intravenous infusion of 0.5% solutions of acetaldehyde into rabbits resulted in the rapid excretion (route not stated) at the rate of 7-10 mg/minute (IARC 1985). N-nitroso-2-methylthiazolidine 4-carboxylic acid was detected in the urine of human subjects during oral and nasal breathing (ACGIH 1991).

B. Acute Effects

Humans exposed acutely to moderate concentrations of acetaldehyde experience irritation of the eyes and respiratory tract and altered respiratory function. Animals exposed to moderate to concentrations exhibit skin and eye irritation and notable cellular alterations in the respiratory epithelium and hyperkeratosis of the fore stomach.

1. Humans - The acute irritation of acetaldehyde is characterized by the following: eye irritation in sensitive individuals, at ppm for 15 minutes; eye irritation, at 50

ppm for 15 minutes; irritation of respiratory tract, at 134 ppm for 30 minutes (2.15 mg/kg over 30 minutes) (see end note 1); irritation of nose and throat, 200 ppm at 15 minutes (Verschueren 1983). Intravenous infusion of human subjects with 5% acetaldehyde at a rate of 20.6-82.4 mg/min for up to 36 minutes (the lowest dose converts to 10.6 mg/kg over 36 minutes) resulted in an increased heart rate, increased ventilation rates and respiratory dead space, and a decreased alveolar carbon dioxide level (IARC 1985).

2. Animals - The oral LD50 value for the rat is 1.93 g/kg (Brabec 1981). The inhalation LC50 for rats exposed for 30 minutes was 20,000 ppm (ACGIH 1991). Acetaldehyde elicited mild skin irritation (open test) and severe eye irritation in rabbits (ACGIH 1991). Intravenous doses of 1-40 mg/kg had a sympathomimetic effect on the heart; doses of <20 mg/kg elevated the blood pressure, and doses >20 mg/kg caused bradycardia and hypotension (ACGIH 1991).

C. Subchronic/Chronic Effects

Prolonged dermal exposure to acetaldehyde can cause erythema and burns in humans; repeat contact may result in dermatitis. Repeat by inhalation, at high concentrations, causes adverse respiratory tract effects in animals. EPA has derived an inhalation reference concentration (RfC) (see end note 2) of 0.009 mg/m^3 for acetaldehyde, based on adverse effects, including degeneration of olfactory epithelia, observed in animal studies.

1. Humans - Prolonged contact of acetaldehyde with the skin may result in erythema and burns; repeated contact with the skin may cause dermatitis, resulting from irritation or sensitization (IARC 1985).
2. Animals - Two 4-week animal studies were the basis for the derivation of the inhalation reference concentration (RfC) for acetaldehyde (U.S. EPA 1994). In the first study, Wistar rats were exposed to 0, 150, or 500 ppm acetaldehyde 6 h/day, 5 days/week for 4 weeks. At the end of the study, lavage was collected

> from the lung and the cells were evaluated; from animals exposed to 500 ppm exhibited decreased viability, decreased cell density, and decreased phagocytosis. Microscopic examination revealed degeneration of the olfactory epithelium. No compound-related effects were observed in rats exposed to 150 ppm [identified as a no-observable-adverse-effect level (NOAEL)]. In the second study, Wistar rats were exposed to 0, 400, 1000, 2200, or 5000 ppm acetaldehyde 6 h/day, 5 days/week for 4 weeks. Effects observed included: death (2200 and 5000 ppm); decreased percentage of lymphocytes; increased number of neutrophilic leukocytes; and decreased organ weights (5000 ppm); growth retardation (three highest concentrations); dose-related degeneration of the nasal olfactory epithelium (all concentrations); and dose-related hyper- and metaplastic changes of the olfactory, laryngeal, and tracheal epithelium (1000, 2200 and 5000 ppm). The 400-ppm concentration, the lowest dose tested, was identified as a lowest-observable adverse-effect level (LOAEL).

The above studies were selected for the determination of the RfC because the same types of lesions appear at longer exposure times and higher exposure levels in chronic studies, and they were concentration (dose)-related. The 150 ppm value from the first study was established as the NOAEL, and the 400 ppm value from the second study was established as the LOAEL. The RfC is 0.009 mg/m^3 (U.S. EPA 1994).

Wistar rats inhaled acetaldehyde concentrations of 0, 750,1500 and 3000/1000 ppm for 6 h/day, 5 days/week for up to 28 months (U.S. EPA 1994). Early mortality in the 3000-ppm group prompted the reduction of the concentration to 1000 ppm. All rats in this group were dead by 25 months, and all but a few were dead in the 1500-ppm group; the cause of early death or moribund condition was almost always partial or complete occlusion of the nose by excessive amounts of keratin and inflammatory exudate. The incidence of basal cell hyperplasia of the olfactory epithelium, increased in low and mid-dose rats, was lower in the high-dose group because of the increased incidence of adenocarcinomas (see section IV.D). Hyperplasia and squamous metaplasia, occasionally with keratinization, developed in the larynx of rats exposed to 1500 and 3000/1000 ppm. A LOAEL of 750 ppm was identified for this study (U.S. EPA 1994). In another study, rats exposed to the same concentrations for 52 weeks were allowed to recover for 26 or 52 weeks. The study demonstrated incomplete recovery of the olfactory and respiratory epithelium even after 52 weeks of recovery (U.S. EPA 1994).

Hamsters exposed by inhalation to acetaldehyde concentrations of 0, 390, 1340, or 4560 ppm for 6 h/day, 5 days/week for 90 days, exhibited decreased body weights; increased organ weights; and effects on the respiratory epithelium that included necrosis, inflammation, hyperplasia and metaplasia (at 4560 ppm); and statistically significant increase in kidney weight in males and small areas of stratified epithelium in the trachea (at 1340 ppm). The authors identified 390 ppm as the NOAEL for the study (U.S. EPA 1994).

Male and female rats received acetaldehyde in the drinking water (25, 125, or 675 mg/kg/day for 4 weeks (HSDB 1994). The only reported adverse effect, hyperkeratosis of the fore stomach, was observed at the high dose.

D. Carcinogenicity

There is inadequate evidence for the carcinogenicity ofacetaldehyde for humans. Based on the carcinogenicity of acetaldehyde in animals, EPA has concluded that acetaldehyde is a probable human carcinogen.

1. Humans - Workers in an aldehyde factory in the German Democratic Republic had a higher cancer rate than expected in that section of Germany (IARC 1985). The plant's main process was the dimerization of acetaldehyde. Of nine cases of malignant neoplasm, 5 were bronchial tumors and two were carcinomas of the oral cavity. Statistical analyses were not available. All the cases were smokers and other chemicals (e.g., acetaldol, butyraldehyde, crotonaldehyde, other higher aldehydes, and acrolein) were present in varying proportions.
2. Animals - Wistar rats inhaled acetaldehyde concentrations of 0, 750, 1500 and 3000/1000 ppm (0, 1350, 2700,

or 5400/1800 mg/m^3) for 6 h/day, 5 days/week for up to 28 months (U.S. EPA 1994). Early mortality in the 3000-ppm group prompted the reduction of the concentration to 1000 ppm. Adenocarcinomas occurred at all exposure levels, and squamous cell carcinoma of the nasal mucosa occurred at the mid and high concentrations (tumor incidence data were not reported) (U.S. EPA 1994). Hamsters (36/sex/group) inhaled decreasing concentrations of acetaldehyde [2500 ppm (4500 mg/m^3) to 1650 ppm (2970 mg/m^3)] 7 h/day, 5 days/week for 52 weeks (IARC 1985). Six animals, killed at the end of exposure, had no tumors; among the remainder of the animals, killed at 81 weeks, the incidences of laryngeal tumors were increased ($p<0.05$; 0/30, 8/29, 0/28, 5/29 for control males, exposed males, control females, exposed females, respectively).

EPA classifies acetaldehyde as B2, "probable human carcinogen," based on the increased incidences of nasal tumors in male and female rats and laryngeal tumors in male and female hamsters (U.S. EPA 1994). IARC has also concluded that there is sufficient evidence for the carcinogenicity of acetaldehyde to experimental animals and inadequate evidence for its carcinogenicity to humans, resulting in an classification of 2B, "possibly carcinogenic to humans" (IARC 1985).

E. Genotoxicity

In the EPA GENETOX Program, acetaldehyde is reported as positivefor DNA repair in Escherichia coli pol A (Rec-assay) and forsister chromatid exchanges, in vitro, in human lymphocytes andnonhuman cells (GENETOX 1994). Acetaldehyde was not mutagenic inSalmonella or in E. coli, but was positive for chromosomeaberrations and sister chromatid exchange both in vitro and invivo mammalian assays (USEPA 1994).

F. Developmental/Reproductive Toxicity

No information was found for the developmental/reproductive effects of acetaldehyde in humans. Limited evidence from i.p.

studies indicate that acetaldehyde causes adverse developmental effects in animals.

1. Humans - No information was found in the secondary sources searched to indicate that acetaldehyde is a developmental/reproductive toxicant in humans.
2. Animals - Pregnant CF rats were given intraperitoneal (i.p.) doses of 50 mg acetaldehyde/kg/day on gestation days 8-15; the fetuses, collected on gestation days 16 through 21, had delayed ossification and skeletal malformations (e.g., wavy ribs) (HSDB 1994). In another study, pregnant rats received i.p. of 50-100 mg/kg of acetaldehyde on day 10, 11, or 12 of gestation; the fetuses were examined on day 21. Acetaldehyde produced a significant increase in fetal resorptions, growth retardation, delayed skeletogenesis, and an increase in skeletal malformations. When treatment occurred on days 8-15 of gestation, growth retardation and malformations of the face and limbs, and delayed ossification were observed (ACGIH 1991).

G. Neurotoxicity

The neurotoxicity potential of acetaldehyde in humans cannot be determined from available information. Limited evidence indicates that acetaldehyde cause neurological effects in animals, including central nervous system depression and neural degeneration.

1. Humans - The irritant effects of acetaldehyde vapor, suchas coughing and burning of the nose, throat and eyes, usually prevents exposure to a level sufficient to cause depression of the central nervous system (IARC 1985). The results of one study in human volunteers indicated that acetaldehyde penetrates the human blood-cerebrospinal fluid barrier (HSDB 1994).
2. Animals - Rats exposed to LC50 concentrations of acetaldehyde exhibited excitement (ACGIH 1991). Dogs exposed to levels of >134 ppm for 30 minutes

experienced inhibition of the central nervous system, and subsequent decrease in the respiratory rate (HSDB 1994). A single i.p. injection (dose not reported) of acetaldehyde produced sustained neural degeneration in the cerebral cortex of rats (HSDB 1994).

V. ENVIRONMENTAL EFFECTS

A. Toxicity to Aquatic Organisms

Acetaldehyde has moderate acute toxicity to aquatic organisms; acute toxicity values are in the range of greater than 1 mg/L to 100 mg/L. LC50 values reported in AQUIRE 1994 for fish include: 37.2 mg/L for 96 hours, Pimephales promelas, the fathead minnow;and 53 mg/L for 96 hours, Lepomis macrochirus, the bluegill. The EC50 value for the immobilization of Daphnia magna (water flea) is 48 mg/L for 48 hours (AQUIRE 1994).

B. Toxicity to Terrestrial Organisms

No information was found in the secondary sources searched regarding the toxicity of acetaldehyde to terrestrial animals; however, information from laboratory animal studies indicate that adverse effects in animals could occur in the presence of high concentrations of acetaldehyde.

C. Abiotic Effects

Limited information on the abiotic effects of acetaldehyde was found in the secondary sources searched. According to the definition provided in the Federal Register (1992), acetaldehyde is a volatile organic compound (VOC) substance. As a VOC, acetaldehyde can contribute to the formation of photochemical smog in the presence of other VOCs. Potential environmental degradation products of acetaldehyde such as PAN, methyl nitrate, and methyl nitrite are components of photochemical smog.

VI. EPA/OTHER FEDERAL/OTHER GROUP ACTIVITY

The Clean Air Act Amendments of 1990 list acetaldehyde as a hazardous air pollutant. The Agency has enacted restrictions for certain waste streams containing the chemical. Occupational exposure to acetaldehyde is regulated by the Occupational Safety and Health Administration. The permissible exposure limit (PEL) is 200 parts per million parts of air (ppm) (29 CFR 1910.1000).

Federal agencies and other groups that can provide additional information on acetaldehyde are listed in Tables 3 and 4.

Table 3 Epa Offices And Contact Numbers For Information On Acetaldehyde.

EPA OFFICE	LAW	PHONE NUMBER
Pollution & Toxics	Prevention Toxic Substances Control Act	
	(sec. 8A/8D/8E)	(202) 554-1404
	Emergency Planning and Community	
	Right-to-Know Act (EPCRA)	
	Regulations (Sec. 313)	(800) 424-9346
	Toxics Release Inventory data	(202) 260-1531
Air	Clean Air Act	(919) 541-0888
Solid Waste &	Comprehensive Environmental	
Emergency Response	Response, Compensation, and	

(*Continued*)

EPA OFFICE	LAW	PHONE NUMBER
	Liability Act (Superfund)/	
	Resource Conservation and Recovery	
	Act / EPCRA (Sec. 304/311/312)	(800) 424-9346
Water	Clean Water Act	(202) 260-7588

Table 4 Other Federal Offices/Other Group Contact Numbers For Information On Acetaldehyde.

Other Agency/ Department/Other Group	Contact Number
American Conference of Governmental Industrial Hygienists (Recommended Highest Exposure Limit (see end note 3): 25 ppm)	(513) 742-2020
National Institute for Occupational Safety & Health (Recommended Exposure Limit (see end note 4): 18 ppm; Limit of Quantitation)	(800) 356-4674
Occupational Safety & Health Administration (Permissible Exposure Limit	

Other Agency/ Department/Other Group	Contact Number
(see end note 5): 200 ppm) (Check local phone book for phone number under Department of Labor)	

VII. END NOTES

1. Calculated using a factor of 1.8 to convert 134 ppm to 241 mg/m^3 which is multiplied by 0.009 (the calculated 30-minute occupational breathing rate of 0.625 m^3 divided by the assumed adult body weight, 70 kg) to obtain the dose in mg/kg (U.S. EPA 1988).
2. The RfC is an estimate (with uncertainty spanning perhaps an order of magnitude) of the exposure level for the human population, including sensitive subpopulations, that is likely to be without an appreciable risk of deleterious effects during the time period of concern.
3. This is a ceiling exposure limit value that should not be exceeded at any time.
4. Exposure should be reduced to the lowest feasible limit; use of only the most reliable and protective respirators is recommended.
5. The OSHA exposure limit is a time-weighted-average (TWA) concentration that must not be exceeded during any 8-hour workshift of a 40-hour work- week.

CITED REFERENCES

ACGIH. 1991. Documentation of the Threshold Limit Values and BiologicalExposure Indicies, 6th ed. American Conference of Governmental Industrial Hygienists, Inc., Cincinnati, OH, pp. 1-5.

AQUIRE. 1994. EPA ERL-Duluth's Aquatic Ecotoxicology Data Systems. U.S. EPA, Duluth, MN. Retrieved July 1994.

Brabec MJ. 1981. Aldehydes and acetals. In: Clayton GD, Clayton FE, Eds. Patty's Industrial Hygiene and Toxicology, 3rd ed., Vol. 2A. New York: John Wiley & Sons, pp. 2629-2639.

Budavari S, O'Neil MJ, Smith A, Heckelman PE (Eds.). 1989. The Merck Index, 11th ed. Merck & Co., Inc., Rahway, NJ, p. 38.

CHEMFATE. 1994. Syracuse Research Corporation's Environmental Fate Data Bases. Syracuse Research Corporation, Syracuse, NY.

Federal Register. 1992. Part 51 - Requirements for Preparation, Adoption, and Submittal of Implementation Plans. Fed. Reg. 57:3945.

GENETOX. 1994. U.S. EPA GENETOX Program, computerized database. Retrieved 7/94.

HSDB. 1994. Hazardous Substances Data Bank. MEDLARS Online Information Retrieval System, National Library of Medicine. Retrieved June, 1994.

IARC. 1985. Acetaldehyde. In: International Agency for Research on Cancer. IARC Monographs on the Evaluation of Carcinogenic Risk of Chemicals to Man, vol. 36. IARC, Lyon, pp.101-132.

Mannsville. 1990. Chemical Products Synopsis, Acetaldehyde. Mannsville Chemical Products Corporation. January 1990.

Mannsville. 1993. Chemical Products Synopsis, Acetaldehyde. Mannsville Chemical Products Corporation. January 1993.

TRI92. 1994. 1992 Toxic Chemical Release Inventory. Office of Pollution Prevention and Toxics, U.S. Environmental Protection Agency, Washington, D.C.

U.S. EPA. 1988. U.S. Environmental Protection Agency. Methodology for Evaluating Potential Carcinogenicity in Support of Reportable Quantity Adjustments Pursuant to CERCLA Section 102. Carcinogen Assessment Group, Office of Health and Environmental Assessment, U.S. EPA, Washington, D.C., pp. 21, 22. OHEA-C-073.

U.S. EPA. 1994. U.S. Environmental Protection Agency. Integrated Risk Information System (IRIS) Online. Coversheet for Acetaldehyde. Office of Health and Environmental Assessment, U.S. EPA, Cincinnati, OH, Retrieved 7/94.

Verschueren K. 1983. Handbook of Environmental Data on Organic Chemicals. Van Nostrand Reinhold Co., New York, pp. 139-141.

SOURCES SEARCHED FOR FACT SHEET PREPARATION

AQUIRE. 1994. Aquatic Information Retrieval online data base. Chemical Information Systems, Inc., a subsidiary of Fein-Marquart Assoc.

ATSDR. 1989-1994. Agency for Toxic Substances and Disease Registry. Toxicological Profiles. Chamblee, GA: ATSDR.

Budavari S, O'Neil MJ, Smith A, Heckelman PE (Eds.). 1989. The Merck Index, 11th ed. Rahway, N.J.: Merck & Co., Inc.

CHEMFATE. 1994. Syracuse Research Corporation's Environmental Fate Data Bases. Syracuse Research Corporation, Syracuse, NY.

Clayton GD, Clayton FE, Eds. 1981-1982. Patty's Industrial Hygiene and Toxicology, 3rd ed. New York: John Wiley & Sons.

GENETOX. 1994. U.S. EPA GENETOX Program, computerized database.

HSDB. 1994. Hazardous Substances Data Bank. MEDLARS Online Information Retrieval System, National Library of Medicine.

IARC. 1979-1994. International Agency for Research on Cancer. IARC Monographs on the Evaluation of Carcinogenic Risk of Chemicals to Man. Lyon: IARC.

NIOSH (National Institute for Occupational Safety and Health). 1992. NIOSH Recommendations for Occupational Safety and Health. Compendium of Policy Documents and Statements. Cincinnati, OH: NIOSH.

NTP. 1994. National Toxicology Program. Toxicology and Carcinogenesis

Studies. Tech Rep Ser. NTP. 1994. National Toxicology Program. Management Status Report. Produced from NTP Chemtrack system. April 8, 1994. National Toxicology Program, Research Triangle Park, NC.

OSHA. 1994. Occupational Safety and Health Administration. Table Z-2. Limits for Air Contaminants.

RTECS. 1994. Registry of Toxic Effects of Chemical Substances. MEDLARS Online Information Retrieval System, National Library of Medicine.

U.S. Air Force. 1989. The Installation Restoration Toxicology Guide, Vols. 1-5. Wright-Patterson Air Force Base, OH.

U.S. EPA (U.S. Environmental Protection Agency). 1991. Table 302.4 List of Hazardous Substances and Reportable Quantities 40 CFR, part 302.4:3-271.

U.S. EPA. Most current. Drinking Water Regulations and Health Advisories.

Office of Drinking Water, U.S. Environmental Protection Agency, Washington, D.C.

U.S. EPA. Most Current. Health Effects Assessment Summary Tables. Cincinnati, OH: Environmental Criteria and Assessment Office, U.S.EPA.

U.S. EPA reviews such as Health and Environmental Effects Documents, Health and Environmental Effect Profiles, and Health and Environmental Assessments.

U.S. EPA. 1994. Integrated Risk Information System (IRIS) Online. Cincinnati, OH: Office of Health and Environmental Assessment.

Acetic acid, solution, more than 10% but not more than 80% acid	
• Methanecarboxylic acid • Ethanoic acid • Ethylic acid	
Formula	CH_3COOH
Structure	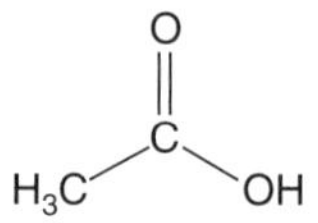
Description	Clear, colorless liquid, odor of vinegar.
Uses	Herbicide, microbiocide, fungicide, ph adjustment.

Registry Numbers and Inventories	
CAS	64-19-7
NIH PubChem CID	176
EC (EINECS/ELINCS)	200-580-7
EC Index Number	607-002-00-6
EC Class	R10, C; R35
RTECS	AF1225000
RTECS class	Mutagen; Reproductive Effector; Human Data; Primary Irritant
UN (DOT)	2790
Merck	13,56
Beilstein/Gmelin	506007
Beilstein Reference	4-02-00-00094
EPA OPP	44001
FEMA	2006
Swiss Giftliste 1	G-4595

Registry Numbers and Inventories	
Canada DSL/NDSL	DSL
US TSCA	Listed
Austrailia AICS	Listed
New Zealand	Listed
Japan ENCS (MITI)	Listed
Korea ECL	Listed
Philippiens PICCS	Listed
Israel	Listed

Properties	
Formula	$C_2H_4O_2$
Formula mass	60.05
Melting point, °C	17
Boiling point, °C	117.9
Vapor pressure, mm_{Hg}	16 (25 C)
Vapor density (air=1)	2.07
Saturation Concentration	1.5% (15000 ppm) at 20 C (calculated)
Evaporization number	24 (diethyl ether = 1)
Odor threshold	0.21 - 1.0 ppm
Critical temperature	321.6
Critical pressure	57.15
Density	1.05 g/cm^3
Solubility in water	Miscible
Viscosity	1.06 cp (25 C)
Surface tension	28.8 g/s2 @ 10 C
Refractive index	1.3718

(*Continued*)

Properties	
Dipole moment	1.5 D
Dielectric constant	6.1
pKa/pKb	4.76 (pKa)
Partition coefficient, pK_{ow}	-0.17
Heat of fusion	11.7 kJ/mol
Heat of vaporization	52.3 kJ/mol
Heat of combustion	-873 kJ/mol

Hazards and Protection	
Storage	Keep away from heat, sparks, and flame. Keep from contact with oxidizing materials. Store in a cool, dry, well-ventilated area away from incompatible substances. Corrosives area. Store in a suitable container in a dry area above the substance's freezing point. Do not store near alkaline substances.
WHMIS	D2B
Handling	Wash thoroughly after handling. Remove contaminated clothing and wash before reuse. Use with adequate ventilation. Empty containers retain product residue, (liquid and/or vapor), and can be dangerous. Avoid contact with heat, sparks and flame. Do not get on skin or in eyes. Avoid ingestion and inhalation. Discard contaminated shoes. Do not pressurize, cut, weld, braze, solder, drill, grind, or expose empty containers to heat, sparks or open flames.

Hazards and Protection	
Protection	Eyes: Wear appropriate protective eyeglasses or chemical safety goggles as described by OSHA's eye and face protection regulations in 29 CFR 1910.133 or European Standard EN166. Skin: Wear appropriate gloves to prevent skin exposure. Clothing: Wear appropriate protective clothing to prevent skin exposure.
Respirators	Follow the OSHA respirator regulations found in 29CFR 1910.134 or European Standard EN 149. Always use a NIOSH or European Standard EN 149 approved respirator when necessary.
Small spills/leaks	Use water spray to dilute spill to a non-flammable mixture. Avoid runoff into storm sewers and ditches which lead to waterways. Wash area with soap and water. Use water spray to disperse the gas/vapor. Remove all sources of ignition. Use a spark-proof tool. Provide ventilation. Cover with material such as dry soda ash or calcium carbonate and place into a closed container for disposal. A vapor suppressing foam may be used to reduce vapors.
Stability	Stable at room temperature in closed containers under normal storage and handling conditions.

(Continued)

Hazards and Protection	
Incompatibilities	Acetaldehyde, 2-aminoethanol, ammonium nitrate, bromine pentafluoride, chlorine trifluoride, chlorosulfonic acid, chromic acid, chromic anhydride + acetic anhydride, diallyl methyl carbinol + ozone, ethylene diamine, ethyleneimine, hydrogen peroxide, nitric acid, nitric acid + acetone, oleum, perchloric acid, permanganates, phosphorus isocyanate, phosphorus trichloride, potassium hydroxide, potassium-t-butoxide, sodium hydroxide, sodium peroxide, and xylene. See NFPA Fire Protection Guide for specifics.
Decomposition	Carbon monoxide, irritating and toxic fumes and gases, carbon dioxide.

Fire	
Flash Point,°C	39
Autoignition, °C	485
Upper exp. limit, %	19.9
Lower exp. limit, %	4
Fire fighting	Use water in flooding quantities as fog. Solid streams of water may be ineffective. Cool all affected containers with flooding quantities of water. Apply water from as far a distance as possible. Use alcohol foam, dry chemical or carbon dioxide. Use water spray to knock-down vapors.
Fire potential	Moderately flammable. Will support combustion.

Hazards		Vapors may form explosive mixtures with air. Vapors may travel to source of ignition and flash back. Most vapors are heavier than air. They will spread along ground and collect in low or confined areas (sewers, basements, tanks). Vapor explosion hazard indoors, outdoors or in sewers.
Combustion products		Irritating vapor generated when heated.
NFPA	Health	3
	Flammability	2
	Reactivity	0

Health	
Exposure limit(s)	OSHA PEL: TWA 10 ppm (25 mg/m^3) NIOSH REL: TWA 10 ppm (25 mg/m^3) ST 15 ppm (37 mg/m^3) NIOSH IDLH: 50 ppm
Poison_Class	4
Exposure effects	Prolonged or repeated skin contact may cause dermatitis. Repeated exposure may lead to blackening and hyperkeratosis of the skin and hands, conjunctivitis, bronchitis and pharyngitis and erosion of the teeth.
Ingestion	May cause severe and permanent damage to the digestive tract. Causes severe pain, nausea, vomiting, diarrhea, and shock. May cause polyuria, oliguria and anuria.
Inhalation	Effects may be delayed. Causes chemical burns to the respiratory tract. Exposure may lead to bronchitis, pharyngitis, and dental erosion.

(Continued)

Health	
Skin	Causes skin burns. May cause skin sensitization, an allergic reaction, which becomes evident upon re-exposure to this material. May be harmful if absorbed through the skin. Contact with the skin may cause blackening and hyperkeratosis of the skin of the hands.
Eyes	Causes severe eye irritation. Contact with liquid or vapor causes severe burns and possible irreversible eye damage. Lachrymator.

First aid	
Ingestion	Do NOT induce vomiting. If victim is conscious and alert, give 2-4 cupfuls of milk or water. Never give anything by mouth to an unconscious person. Get medical aid immediately.
Inhalation	Get medical aid immediately. Remove from exposure to fresh air immediately. If breathing is difficult, give oxygen. DO NOT use mouth-to-mouth respiration. If breathing has ceased apply artificial respiration using oxygen and a suitable mechanical device such as a bag and a mask.
Skin	Get medical aid immediately. Immediately flush skin with plenty of soap and water for at least 15 minutes while removing contaminated clothing and shoes. Wash clothing before reuse. Destroy contaminated shoes.
Eyes	Get medical aid immediately. Do NOT allow victim to rub or keep eyes closed. Extensive irrigation is required (at least 30 minutes).

Transportation		
UN number	2790	
Response guide	153	
Hazard class	8	
USCG CHRIS Code	AAC	
USCG Compatatibility Group	4 Organic acids	
HS Code	2915 21 00	
Std. Transport #	4931401	
IMO Chemical Code	17	
IMO Pollution Category	D	
IMO Hazard code	S	

Ammonium persulfate	
• Ammonium peroxydisulfate • Diammonium peroxodisulfate • Diazanium sulfonatooxy sulfate	
Formula	$(NH_4)_2S_2O_8$
Structure	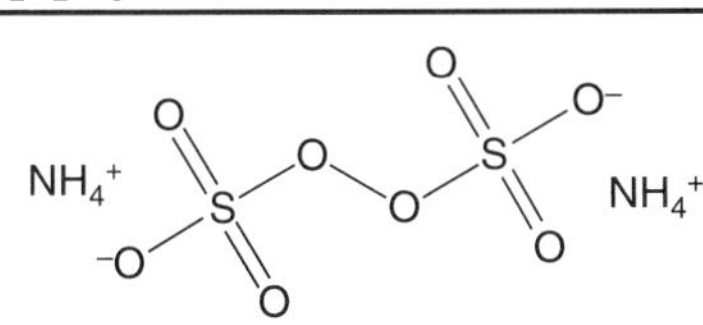
Description	White crystals, odorless.
Uses	Used as an oxidizer and bleacher.

Registry Numbers and Inventories	
CAS	7727-54-0
NIH PubChem CID	62648
EC (EINECS/ELINCS)	231-786-5
EC Index Number	016-060-00-6
EC Class	O; R8, Xn; R22, Xi; R36/37/38, R42/43
RTECS	SE0350000
RTECS class	Other
UN (DOT)	1444
Merck	13,545
Beilstein/Gmelin	47998 (G)
Swiss Giftliste 1	G-1115
Canada DSL/NDSL	DSL
US TSCA	Listed
Austrailia AICS	Listed
New Zealand	Listed
Japan ENCS (MITI)	Listed
Korea ECL	Listed
Philippiens PICCS	Listed

Properties	
Formula	H8N2O8S2
Formula mass	228.20
Melting point, °C	120
Decomposition point, °C	120
Vapor density (air=1)	7.9
Density	1.98 g/cm^3 (20 C)
Solubility in water	800 g/L

Hazards and Protection	
Storage	Do not store near combustible materials. Store in a tightly closed container. Store in a cool, dry, well-ventilated area away from incompatible substances. Keep away from metals. Keep away from reducing agents.
Handling	Wash thoroughly after handling. Use with adequate ventilation. Minimize dust generation and accumulation. Avoid contact with eyes, skin, and clothing. Keep container tightly closed. Avoid contact with clothing and other combustible materials. Avoid ingestion and inhalation. Wash clothing before reuse.
Protection	Eyes: Wear appropriate protective eyeglasses or chemical safety goggles as described by OSHA's eye and face protection regulations in 29 CFR 1910.133 or European Standard EN166. Skin: Wear appropriate gloves to prevent skin exposure. Clothing: Wear appropriate protective clothing to minimize contact with skin.
Respirators	Follow the OSHA respirator regulations found in 29CFR 1910.134 or European Standard EN 149. Always use a NIOSH or European Standard EN 149 approved respirator when necessary.
Small spills/leaks	Avoid runoff into storm sewers and ditches which lead to waterways. Clean up spills immediately, using the appropriate protective equipment. Sweep up, then place into a suitable container for disposal. Provide ventilation.
Stability	Substance is shock sensitive and thermally unstable. Decomposes when heated. May decompose on exposure to moist air or water.

(*Continued*)

Hazards and Protection	
Incompatibilities	Reducing agents, iron, sodium peroxide, powdered aluminum.
Decomposition	Nitrogen oxides, oxides of sulfur, ammonia and/or derivatives, nitrogen.

Fire		
Flash Point,°C		80
Fire fighting		Wear a self-contained breathing apparatus in pressure-demand, MSHA/NIOSH (approved or equivalent), and full protective gear. Use water spray to keep fire-exposed containers cool. Oxidizer. Greatly increases the burning rate of combustible materials. This material in sufficient quantity and reduced particle size is capable of creating a dust explosion. May accelerate burning if involved in a fire. Containers may explode when heated. Extinguishing media: Cool containers with flooding quantities of water until well after fire is out. For small fires DO NOT use dry chemicals, carbon dioxide, halon or foams. USE WATER ONLY. For large fires, flood fire area with water from a distance.
Fire potential		Moderate fire hazard by reaction with reducing agents.
Hazards		May explode from heat or contamination. May react explosively with hydrocarbons (fuels). May ignite combustibles.
Combustion products		Toxic oxides of nitrogen and sulfuric acid fumes may form in fire.
NFPA	**Health**	2

Fire		
	Flammability	1
	Reactivity	1
	Special	O

Health	
Exposure limit(s)	OEL-UNITED KINGDOM:TWA 1 mg/m^3
Poison_Class	4
Exposure effects	Repeated or prolonged exposure may cause allergic reactions in sensitive individuals.
Ingestion	Causes gastrointestinal irritation with nausea, vomiting and diarrhea. May be harmful if swallowed.
Inhalation	Causes respiratory tract irritation. Inhalation may be fatal as a result of spasm, inflammation, edema of the larynx and bronchi, chemical pneumonitis and pulmonary edema.
Skin	Causes skin irritation. May cause skin sensitization, an allergic reaction, which becomes evident upon re-exposure to this material.
Eyes	Causes eye irritation.

First aid	
Ingestion	Do NOT induce vomiting. If victim is conscious and alert, give 2-4 cupfuls of milk or water. Never give anything by mouth to an unconscious person. Get medical aid.

(*Continued*)

First aid	
Inhalation	Remove from exposure to fresh air immediately. If breathing is difficult, give oxygen. Get medical aid if cough or other symptoms appear. DO NOT use mouth-to-mouth respiration. If breathing has ceased apply artificial respiration using oxygen and a suitable mechanical device such as a bag and a mask.
Skin	Flush skin with plenty of soap and water for at least 15 minutes while removing contaminated clothing and shoes. Get medical aid if irritation develops or persists. Wash clothing before reuse.
Eyes	Immediately flush eyes with plenty of water for at least 15 minutes, occasionally lifting the upper and lower eyelids. Get medical aid.

Transportation		
UN number	1444	
Response guide	140	
Hazard class	5.1	
Packing Group	III	
USCG CHRIS Code	APE	
HS Code	2833 40 00	
Std. Transport #	4918706	

Boric acid	
• Hydrogen orthoborate • Orthoboric acid • Boron trihydroxide • Trihydroxyborane • Borofax • Terminator DF	
Formula	H_3BO_3
Structure	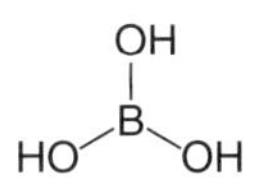
Description	colorless, odorless scales or white powder.
Uses	Weatherproofing wood, nickling baths, for printing & dyeing, & for impregnating wicks, manufacture cements, crockery, porcelain, enamels, glass, borates, leather, carpets, hats, soaps, artificial gems, in painting, photography, electric condensers, hardening steel.

Registry Numbers and Inventories	
CAS	10043-35-3
NIH PubChem CID	7628
EC (EINECS/ELINCS)	233-139-2
EC Index Number	005-007-00-2
EC Class	R: 37, S: 22-24/25
RTECS	ED4550000

(Continued)

Registry Numbers and Inventories	
RTECS class	Agricultural Chemical and Pesticide; Tumorigen; Drug; Mutagen; Reproductive Effector; Human Data; Primary Irritant
Merck	13,1326
Beilstein/Gmelin	1585 (G)
EPA OPP	11001
Swiss Giftliste 1	G-1285
Canada DSL/NDSL	DSL
US TSCA	Listed
Austrailia AICS	Listed
New Zealand	Listed
Japan ENCS (MITI)	Listed
Korea ECL	Listed
Philippiens PICCS	Listed

Properties	
Formula	BH3O3
Formula mass	61.83
Melting point, °C	185
Decomposition point, °C	300
Vapor pressure, mm_{Hg}	2 (20 C)
Odor threshold	Odorless
Density	1.435 g/cm^3 (15 C)
Solubility in water	49 g/L
Refractive index	1.462
pKa/pKb	8.91 (pKa)
Heat of vaporization	98.1 kJ/mol

Hazards and Protection	
Storage	Store in a cool, dry, well-ventilated area away from incompatible substances. Keep containers tightly closed.
Handling	Wash thoroughly after handling. Use only in a well ventilated area. Minimize dust generation and accumulation. Avoid contact with eyes, skin, and clothing. Keep container tightly closed. Avoid ingestion and inhalation.
Protection	Eyes: Wear appropriate protective eyeglasses or chemical safety goggles as described by OSHA's eye and face protection regulations in 29 CFR 1910.133 or European Standard EN166. Skin: Wear appropriate protective gloves to prevent skin exposure. Clothing: Wear appropriate protective clothing to prevent skin exposure.
Respirators	Follow the OSHA respirator regulations found in 29CFR 1910.134 or European Standard EN 149. Always use a NIOSH or European Standard EN 149 approved respirator when necessary.
Small spills/leaks	Vacuum or sweep up material and place into a suitable disposal container. Avoid generating dusty conditions. Provide ventilation.
Stability	Stable at room temperature in closed containers under normal storage and handling conditions.
Incompatibilities	Incompatible with acetic anhydride and potassium. Reacts with basic materials to form borate salts.
Decomposition	Oxides of boron.

Fire		
Fire fighting		Wear a self-contained breathing apparatus in pressure-demand, MSHA/NIOSH (approved or equivalent), and full protective gear. Extinguishing media: Substance is noncombustible; use agent most appropriate to extinguish surrounding fire.
Fire potential		Not flammable
NFPA	**Health**	1
	Flammability	0
	Reactivity	0

Health	
Exposure limit(s)	OEL-RUSSIA:STEL 10 mg/m^3
Poison_Class	4
Exposure effects	Prolonged or repeated skin contact may cause dermatitis. May cause borism characterized by dry skin, skin eruptions, and gastric disturbances.. Infants and young children are more susceptible to boric poisoning.
Ingestion	Causes gastrointestinal irritation with nausea, vomiting and diarrhea. May cause central nervous system depression, characterized by excitement, followed by headache, dizziness, drowsiness, and nausea. Advanced stages may cause collapse, unconsciousness, coma and possible death due to respiratory failure. May cause tissue anoxia, characterized by weakness, headache, dizziness, confusion, cyanosis, weak and irregular heart beat, collapse, unconsciousness, convulsions, coma and death.
Inhalation	May cause respiratory tract irritation.

Health	
Skin	May cause skin irritation. May be absorbed through damaged or abraded skin in harmful amounts.
Eyes	May cause eye irritation.
First aid	
Ingestion	Induce vomiting. If victim is conscious and alert, give 2-4 cupfuls of milk or water. Get medical aid immediately.
Inhalation	Get medical aid immediately. Remove from exposure to fresh air immediately. If not breathing, give artificial respiration. If breathing is difficult, give oxygen.
Skin	Flush skin with plenty of soap and water for at least 15 minutes while removing contaminated clothing and shoes. Get medical aid if irritation develops or persists. Wash clothing before reuse.
Eyes	Flush eyes with plenty of water for at least 15 minutes, occasionally lifting the upper and lower eyelids. Get medical aid.

Transportation	
USCG CHRIS Code	BAC
HS Code	2810 00 90
IMO Pollution Category	D

Calcium Chloride	
• Hydrophilite • Calcosan • Dowflake • Liquidow	
Formula	$CaCl_2$
Structure	Ca^{2+} Cl^- Cl^-
Description	White, crystalline solid. Odorless. Solutions are clear to light yellow. Hygroscopic or deliquescent.
Uses	Pavement deicer.

Registry Numbers and Inventories	
CAS	10043-52-4
NIH PubChem CID	24854
EC (EINECS/ELINCS)	233-140-8
EC Index Number	017-013-00-2
EC Class	Xi; R36
RTECS	EV9800000
RTECS class	Tumorigen; Mutagen; Human Data
Merck	13,1660
Beilstein/Gmelin	17127 (G)
EPA OPP	75605
Swiss Giftliste 1	G-4896
Canada DSL/NDSL	DSL
US TSCA	Listed
Austrailia AICS	Listed

Registry Numbers and Inventories	
New Zealand	Listed
Japan ENCS (MITI)	Listed
Korea ECL	Listed
Philippiens PICCS	Listed

Properties	
Formula	$CaCl_2$
Formula mass	110.99
Melting point, °C	772
Boiling point, °C	1600
Density	2.174 g/cm^3
Solubility in water	745 g/L
Thermal expansion	0.0000227/K
Heat of fusion	26.31 kJ/mol
Heat of vaporization	235 kJ/mol

Hazards and Protection	
Storage	Store in a tightly closed container. Store in a cool, dry, well-ventilated area away from incompatible substances. Store protected from moisture.
WHMIS	D2B
Handling	Wash thoroughly after handling. Use with adequate ventilation. Minimize dust generation and accumulation. Keep container tightly closed. Do not get on skin or in eyes. Do not ingest or inhale. Wash clothing before reuse.

(*Continued*)

Hazards and Protection	
Protection	Eyes: Wear appropriate protective eyeglasses or chemical safety goggles as described by OSHA's eye and face protection regulations in 29 CFR 1910.133 or European Standard EN166. Skin: Wear impervious gloves. Clothing: Wear appropriate protective clothing to minimize contact with skin.
Respirators	Follow the OSHA respirator regulations found in 29CFR 1910.134 or European Standard EN 149. Always use a NIOSH or European Standard EN 149 approved respirator when necessary.
Small spills/leaks	Vacuum or sweep up material and place into a suitable disposal container. Clean up spills immediately, using the appropriate protective equipment. Avoid generating dusty conditions. Provide ventilation.
Disposal code	14
Stability	Stable under normal shipping and handling conditions.
Incompatibilities	Moisture.
Decomposition	Chlorine.

Fire		
Fire fighting		Fire control: Use water, dry chemical, carbon dioxide or foam to extinguish. Do not extinguish fire unless release can be stopped. Cool fire-exposed containers with water.
Fire potential		Nonflammable.
NFPA	**Health**	2
	Flammability	0
	Reactivity	0

Health	
Poison_Class	-
Exposure effects	Effects may be delayed.
Ingestion	May cause severe gastrointestinal tract irritation with nausea, vomiting and possible burns. May cause cardiac disturbances. May be harmful if swallowed. In very severe cases, seizures, rapid respiration, slow heartbeat, or death, may
Inhalation	May cause severe irritation of the upper respiratory tract with pain, burns, and inflammation.
Skin	Contact with skin causes irritation and possible burns, especially if the skin is wet or moist.
Eyes	Contact with eyes may cause severe irritation, and possible eye burns.

First aid	
Ingestion	Do NOT induce vomiting. If victim is conscious and alert, give 2-4 cupfuls of milk or water. Never give anything by mouth to an unconscious person. Get medical aid.
Inhalation	Remove from exposure to fresh air immediately. If not breathing, give artificial respiration. If breathing is difficult, give oxygen. Get medical aid. DO NOT use mouth-to-mouth respiration.
Skin	Get medical aid. Immediately flush skin with plenty of soap and water for at least 15 minutes while removing contaminated clothing and shoes. Wash clothing before reuse.
Eyes	Immediately flush eyes with plenty of water for at least 15 minutes, occasionally lifting the upper and lower eyelids. Get medical aid.

Transportation	
USCG CHRIS Code	CLC
HS Code	2827 20 00

Citric Acid	
• 2-Hydroxy-1,2,3-propanetricarboxylic acid • 3-Carboxy-3-hydroxypentane-1,5-dioic acid • Hydrocerol A • Citro • Chemfill • Aciletten	
Formula	$(HOOCCH_2)_2C(OH)COOH$
Structure	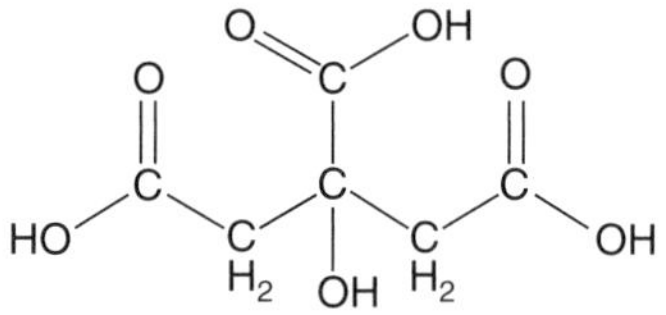
Description	Colorless crystals, or white, granular to fine crystalline powder, deliquescent in moist air.
Uses	Acidulant in beverages, to adjust ph of foods and as a synergistic antioxidant, in processing cheese, used in beverages, jellies, jams, preserves and candy to provide tartness, in special inks, in electroplating, in the manufacture of alkylated resins, in esterified form as plasticizer, foam inhibitor, in the manufacture of citric acid salts, as mordant to brighten colors.

Registry Numbers and Inventories	
CAS	77-92-9
NIH PubChem CID	311
EC (EINECS/ELINCS)	201-069-1
EC Class	Xi, R: 37/38-41, S: 26-36/37/39
RTECS	GE7350000
RTECS class	Agricultural Chemical and Pesticide; Drug; Primary Irritant
UN (DOT)	1759
Merck	13,2350
Beilstein/Gmelin	782061
Beilstein Reference	4-03-00-01272
EPA OPP	21801
FEMA	2306
Swiss Giftliste 1	G-1438
Canada DSL/NDSL	DSL
US TSCA	Listed
Austrailia AICS	Listed
New Zealand	Listed
Japan ENCS (MITI)	Listed
Korea ECL	Listed
Philippiens PICCS	Listed

Properties	
Formula	C6H8O7
Formula mass	192.13
Melting point, °C	149 - 151
Boiling point, °C	310
Vapor pressure, mm_{Hg}	6E-5 (25 C)

(*Continued*)

Properties	
Evaporization number	Probably very low at normal temperatures.
Odor threshold	Odorless
Density	1.665 g/cm^3
Solubility in water	590 g/L
Viscosity	6.5 cp 50% aq soln @ 25C
Refractive index	1.44
pKa/pKb	2.93 (pKa)
Partition coefficient, pK_{ow}	-1.72
Heat of vaporization	63.8 kJ/mol
Heat of combustion	-1983 kJ/mol

Hazards and Protection	
Storage	Store in a tightly closed container. Store in a cool, dry, well-ventilated area away from incompatible substances. Store protected from moisture.
WHMIS	E
Handling	Wash thoroughly after handling. Use with adequate ventilation. Minimize dust generation and accumulation. Avoid contact with eyes, skin, and clothing. Keep container tightly closed. Avoid ingestion and inhalation. Do not allow contact with water. Keep from contact with moist air and steam.
Protection	Eyes: Wear appropriate protective eyeglasses or chemical safety goggles as described by OSHA's eye and face protection regulations in 29 CFR 1910.133 or European Standard EN166. Skin: Wear appropriate protective gloves to prevent skin exposure. Clothing: Wear appropriate protective clothing to prevent skin exposure.

Hazards and Protection	
Respirators	Follow the OSHA respirator regulations found in 29CFR 1910.134 or European Standard EN 149. Always use a NIOSH or European Standard EN 149 approved respirator when necessary.
Small spills/leaks	Vacuum or sweep up material and place into a suitable disposal container. Very fine particles can cause a fire or explosion. Eliminate all ignition sources. Clean up spills immediately, using the appropriate protective equipment. Avoid generating dusty conditions. Remove all sources of ignition. Provide ventilation. Spill may be neutralized with lime. Do not get water inside containers.
Stability	Stable under normal temperatures and pressures.
Incompatibilities	Oxidizing agents, sulfides (inorganic, e.g. ferric sulfide, lead sulfide, sodium sulfide), metal nitrates, alkali carbonates, alkalis, potassium tartrate, acetates, bicarbonates.
Decomposition	Carbon monoxide, irritating and toxic fumes and gases, carbon dioxide.

Fire	
Flash Point,°C	155
Autoignition, °C	345
Upper exp. limit, %	4.8
Lower exp. limit, %	1.8

(Continued)

Fire		
Fire fighting		Wear a self-contained breathing apparatus in pressure-demand, MSHA/NIOSH (approved or equivalent), and full protective gear. During a fire, irritating and highly toxic gases may be generated by thermal decomposition or combustion. This material in sufficient quantity and reduced particle size is capable of creating a dust explosion. To extinguish fire use water spray, dry chemical, carbon dioxide, or chemical foam. Use agent most appropriate to extinguish fire. Do NOT get water inside containers.
Fire potential		Slight - when heated.
Hazards		Melts and decomposes. The reaction is not hazardous.
Combustion products		Fire may produce irritating, corrosive and/or toxic gases.
NFPA	**Health**	2
	Flammability	0
	Reactivity	0

Health	
Poison_Class	5
Exposure effects	Repeated exposure may cause sensitization dermatitis.
Ingestion	May cause gastrointestinal irritation with nausea, vomiting and diarrhea. Excessive intake of citric acid may cause erosion of the teeth.
Inhalation	Causes respiratory tract irritation.
Skin	Causes skin irritation. May cause skin sensitization, an allergic reaction, which becomes evident upon re-exposure to this material.

Eyes	Causes severe eye irritation and possible injury.
First aid	
Ingestion	Do NOT induce vomiting. If victim is conscious and alert, give 2-4 cupfuls of milk or water. Never give anything by mouth to an unconscious person. Get medical aid.
Inhalation	Remove from exposure to fresh air immediately. If breathing is difficult, give oxygen. Get medical aid. DO NOT use mouth-to-mouth respiration. If breathing has ceased apply artificial respiration using oxygen and a suitable mechanical device such as a bag and a mask.
Skin	Flush skin with plenty of soap and water for at least 15 minutes while removing contaminated clothing and shoes. Get medical aid if irritation develops or persists. Wash clothing before reuse.
Eyes	Immediately flush eyes with plenty of water for at least 15 minutes, occasionally lifting the upper and lower eyelids. Get medical aid. Do NOT allow victim to rub or keep eyes closed.

Transportation		
UN number	1759	
Response guide	154	
Hazard class	8	
Packing Group	I; II; III	
USCG CHRIS Code	CIT	
USCG Compatatibility Group	4 Organic acids	
HS Code	2918 14 00	
IMO Pollution Category	D	

Distillates (petroleum), light distillate, low-boiling	
Description	Clear, water white to pale yellow liquid with a mild petroleum odor.

Registry Numbers and Inventories	
CAS	68410-97-9
NIH PubChem CID	48420467 (SID)
EC (EINECS/ELINCS)	270-093-2
EC Index Number	649-332-00-3
EC Class	Carc. Cat. 2; R45, Xn; R65
Canada DSL/NDSL	DSL
US TSCA	Listed
Austrailia AICS	Listed
New Zealand	Listed
Korea ECL	Listed
Philippiens PICCS	Listed

Properties	
Vapor pressure, mm_{Hg}	518 (20°C)
Vapor density (air=1)	3
Density	0.69 g/cm^3 (20°C)
Solubility in water	Insoluble.

Hazards and Protection	
Storage	Store as OSHA Class IB flammable liquid. Keep away from flames, sparks or hot surfaces. Never use a torch to cut or weld on or near container. Store at ambient temperature or less.
WHMIS	KE-12565
Protection	Safety glasses, splash goggles, of face shield as appropriate. Have suitable eye water wash available. Wear protective impervious gloves and clothing. Acceptable materials for gloves are neoprene; nitrilee; viton.
Respirators	Up to 500 ppm hexane / 4000 ppm naptha, half mask organic vapor respirator. Up to 2500 ppm hexane / 20,000 ppm naptha, full face organic vapor respirator or full face supplied air respirator. Greater that 2500 ppm hexane / 20,000 ppm naptha, fire fighting, or unknown concentration, self contained breathing apparatus with positive pressure.
Small spills/leaks	Notify emergency response personnel. Evacuate area and remove ignition sources. Build dike to contain flow. Remove free liquid. Shut off ignition source. Contain spill and keep from entering waterways or sewers. Use personal protective equipment. Advise EPA; state agency if required. Absorb on inert material. Shovel, sweep or vacuum spill and place in closed container for disposal.
Stability	Stable at normal temperatures and pressures.
Incompatibilities	Strong oxidizing agents.
Decomposition	Incomplete combustion may produce fumes, smoke, carbon monoxide and other asphyxiants.

Fire	
Flash Point,°C	< -18 C
Fire fighting	Extinguish with dry Chemical, carbon dioxide, and foam. CAUTION: Water stream may spread fire. Use water spray only to cool containers exposed to flames. Do not enter enclosed or a confined work space without proper protective equipment. Fire fighting personnel should wear respiratory protection (positive pressure if available). If leak or spill has not ignited, use water spray to disperse the vapors.
Hazards	Can form flammable mixtures with air and flash at room temperature. Explosion hazard in fire situation. Vapor heavier than air and may travel considerable distance to a source of ignition and flash back.
Combustion products	Products of combustion include fumes, smoke and carbon monoxide.

Health	
Exposure effects	
Ingestion	Acute aspiration hazard. Tests on similar materials indicate the possibility of the following symptoms: headache, nausea, drowsiness, fatigue, pneumonitis, pulmonary adema, central nervous system depression, convulsions, and loss of consciousness.
Inhalation	Acute Irritation. Tests on similar materials indicate the possibility of the following symptoms: headache, nasal and respiratory irritation, nausea, drowsiness, breathlessness, fatigue, central nervous system depression, convulsions, and loss of consciousness.

Health	
Skin	Acute irritation upon short-term exposure and chronic dermatitis on prolonged contact.
Eyes	Acute irritation.
First aid	
Ingestion	Call a physician immediately. Do not induce vomiting except at the instruction of a physician. Never give anything by mouth to an unconscious person.
Inhalation	Remove patient to fresh air and consult a physician. If breathing is difficult, give oxygen. If not breathing give artificial respiration.
Skin	Wash skin thoroughly with soap and water. Immediately remove contaminated clothing and launder before reuse. If irritation or rash develops, obtain medical assistance.
Eyes	Flush eyes immediately with water a minimum of 15 minutes occasionally lifting lower and upper lids. Get medical attention promptly.

Ethylene Glycol Monobutyl Ether	
• 2-Butoxyethanol • Butyl cellusolve • Butyl glycol • Butyl oxitol	
Formula	$C_4H_9OCH_2CH_2OH$

(*Continued*)

Ethylene Glycol Monobutyl Ether	
Structure	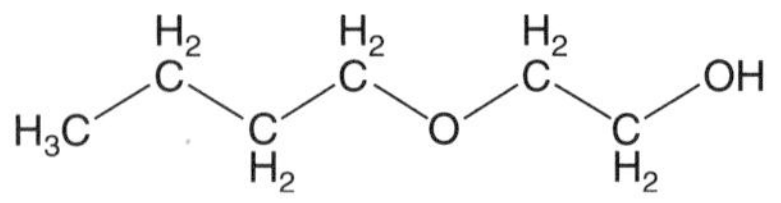
Description	Colorless liquid with a weak, pleasant, sweet, ester odor.
Uses	In hydraulic fluids.

Registry Numbers and Inventories	
CAS	111-76-2
NIH PubChem CID	8133
EC (EINECS/ELINCS)	203-905-0
EC Index Number	603-014-00-0
EC Class	Xn; R20/21/22, Xi; R36/38
RTECS	KJ8575000
RTECS class	Tumorigen; Mutagen; Reproductive Effector; Human Data; Primary Irritant
UN (DOT)	2369
Merck	13,1558
Beilstein/Gmelin	1732511
Beilstein Reference	4-01-00-02380
EPA OPP	11501
Swiss Giftliste 1	G-1334
Canada DSL/NDSL	DSL
US TSCA	Listed
Austrailia AICS	Listed
New Zealand	Listed
Japan ENCS (MITI)	Listed
Korea ECL	Listed
Philippiens PICCS	Listed

Properties	
Formula	C6H14O2
Formula mass	118.18
Melting point, °C	-70
Boiling point, °C	171
Vapor pressure, mm_{Hg}	0.8 (20 C)
Vapor density (air=1)	4.1
Saturation Concentration	0.1% at 20 C
Evaporization number	0.07 (butyl acetate = 1)
Critical temperature	370
Critical pressure	38.50
Density	0.902 g/cm^3 (20 C)
Solubility in water	Miscible
Viscosity	1.87 cp (40 C)
Surface tension	27.4 g/s2 @ 25 C.
Refractive index	1.4196 (20 C)
Dipole moment	1.8 D
Dielectric constant	9.4 (20 C)
Partition coefficient, pK_{ow}	0.83
Heat of vaporization	47.1 kJ/mol
Heat of combustion	-4163 kJ/mol

Hazards and Protection	
Storage	Keep away from heat, sparks, and flame. Keep away from sources of ignition. Store in a tightly closed container. Store in a cool, dry, well-ventilated area away from incompatible substances.
WHMIS	B3 D1A D2B

(Continued)

Hazards and Protection	
Handling	Wash thoroughly after handling. Use only in a well ventilated area. Ground and bond containers when transferring material. Do not get in eyes, on skin, or on clothing. Empty containers retain product residue, (liquid and/or vapor), and can be dangerous. Keep container tightly closed. Avoid contact with heat, sparks and flame. Do not ingest or inhale. Do not pressurize, cut, weld, braze, solder, drill, grind, or expose empty containers to heat, sparks or open flames.
Protection	Eyes: Wear appropriate protective eyeglasses or chemical safety goggles as described by OSHA's eye and face protection regulations in 29 CFR 1910.133 or European Standard EN166. Skin: Wear appropriate protective gloves to prevent skin exposure. Clothing: Wear appropriate protective clothing to prevent skin exposure.
Respirators	A full facepiece respirator with organic vapor cartridge may be worn up to 50 times the exposure limit or the maximum use concentration specified by the appropriate regulatory agency or respirator supplier, whichever is lowest. For emergencies or instances where the exposure levels are not known, use a full-facepiece positive-pressure, air-supplied respirator. This compound possibly exists in both particulate and vapor phase. A particulate (NIOSH type N95 or better) prefilter should be used for the particulate.

Hazards and Protection	
Small spills/leaks	Absorb spill with inert material, (e.g., dry sand or earth), then place into a chemical waste container. Avoid runoff into storm sewers and ditches which lead to waterways. Clean up spills immediately, using the appropriate protective equipment. Remove all sources of ignition. Use a spark-proof tool. Provide ventilation.
Disposal code	1
Stability	Stable under normal temperatures and pressures.
Incompatibilities	Strong oxidizers and strong bases.
Decomposition	Carbon monoxide, irritating and toxic fumes and gases, carbon dioxide.

Fire	
Flash Point,°C	62
Autoignition, °C	238
Upper exp. limit, %	12.7
Lower exp. limit, %	1.1
Fire fighting	Wear full protective clothing and NIOSH-approved self-contained breathing apparatus with full facepiece operated in the pressure demand or other positive pressure mode. Extinguish with dry chemical, alcohol foam or carbon dioxide. Do not use a solid stream of water, since the stream will scatter and spread the fire. Water spray may be used to keep fire exposed containers cool, dilute spills to nonflammable mixtures, protect personnel attempting to stop leak and disperse vapors

(*Continued*)

Fire		
Fire potential		May burn but does not ignite readily.
Hazards		Containers may explode when heated. Runoff may pollute waterways. Substance may be transported in a molten form.
Combustion products		Fire may produce irritating, corrosive and/or toxic gases.
NFPA	**Health**	2
	Flammability	2
	Reactivity	0

Health	
Exposure limit(s)	TLV: 25 ppm; 121 mg/m^3 (as TWA) (skin) (ACGIH 1992-1993) OSHA PEL: TWA 50 ppm (240 mg/m^3) skin NIOSH REL: TWA 5 ppm (24 mg/m^3) skin NIOSH IDLH: 700 ppm
Poison_Class	4
Exposure effects	Prolonged or repeated exposures can cause damage to the liver, kidneys, lymphoid system, blood and blood-forming organs.
Ingestion	Causes irritation to the gastrointestinal tract. Symptoms may include nausea, vomiting and diarrhea. Toxic! May cause systemic poisoning with symptoms paralleling those of inhalation.
Inhalation	Causes irritation to the respiratory tract. Symptoms may include sore throat, coughing, headache, nausea and shortness of breath. High concentrations have a narcotic effect.

Health	
Skin	May cause irritation with redness and pain. May be absorbed through the skin with possible systemic effects.
Eyes	Vapors are irritating and may produce immediate pain, redness and tearing. Splashes can cause severe pain, stinging, swelling.
First aid	
Ingestion	Do NOT induce vomiting. If victim is conscious and alert, give 2-4 cupfuls of milk or water. Never give anything by mouth to an unconscious person. Get medical aid immediately.
Inhalation	Get medical aid immediately. Remove from exposure to fresh air immediately. If not breathing, give artificial respiration. If breathing is difficult, give oxygen. DO NOT use mouth-to-mouth respiration.
Skin	Get medical aid immediately. Immediately flush skin with plenty of soap and water for at least 15 minutes while removing contaminated clothing and shoes. Wash clothing before reuse.
Eyes	Flush eyes with plenty of water for at least 15 minutes, occasionally lifting the upper and lower eyelids. Get medical aid immediately.

Transportation		
UN number	2369	
Response guide	152	
Hazard class	6.1	
Packing Group	III	
USCG CHRIS Code	EGM	
USCG Compatatibility Group	40 Glycol ethers	
HS Code	2909 43 00	
Std. Transport #	4925128 4925229	
IMO Chemical Code	17	
IMO Pollution Category	D	
IMO Hazard code	S	

Ethylene Glycol	
• 1,2-Ethanediol • Ethane-1,2-diol • Monoethylene glycol	
Formula	$HOCH_2CH_2OH$
Structure	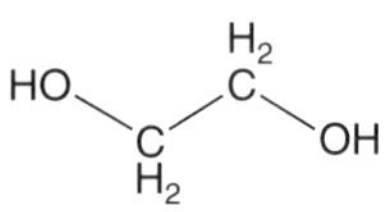
Description	Clear, colorless, syrupy liquid, hygroscopic.
Uses	Automotive antifreeze, solvent, adjuvant.

Registry Numbers and Inventories	
CAS	107-21-1
NIH PubChem CID	174
EC (EINECS/ELINCS)	203-473-3
EC Index Number	603-027-00-1
EC Class	Xn; R22
RTECS	KW2975000
RTECS class	Agricultural Chemical and Pesticide; Tumorigen; Mutagen; Reproductive Effector; Human Data; Primary Irritant
Merck	13,3832
Beilstein/Gmelin	505945
Beilstein Reference	4-01-00-02369
EPA OPP	42203
Swiss Giftliste 1	G-1177
Canada DSL/NDSL	DSL
US TSCA	Listed
Austrailia AICS	Listed
New Zealand	Listed
Japan ENCS (MITI)	Listed
Korea ECL	Listed
Philippiens PICCS	Listed
Israel	Listed

Properties	
Formula	$C_2H_6O_2$
Formula mass	62.07
Melting point, °C	-13
Boiling point, °C	195
Decomposition point, °C	451

(Continued)

Properties	
Vapor pressure, mm$_{Hg}$	0.06 (20 C)
Vapor density (air=1)	2.6
Saturation Concentration	66 ppm at 20 C
Evaporization number	600 (diethyl ether = 1)
Odor threshold	Odorless
Critical temperature	446
Critical pressure	80.9
Density	1.115 g/cm^3 (20 C)
Solubility in water	Miscible
Viscosity	26 cp @ 15C
Surface tension	48.4 g/s2 @ 20 C
Refractive index	1.43312 (20 C)
Dipole moment	2.28 D (20 C)
Dielectric constant	37.7 (25 C)
Partition coefficient, pK$_{ow}$	-1.36
Heat of fusion	11.2 kJ/mol
Heat of vaporization	50.5 kJ/mol
Heat of combustion	-1036 kJ/mol

Hazards and Protection	
Storage	Keep away from sources of ignition. Store in a cool, dry, well-ventilated area away from incompatible substances.
WHMIS	D2A
Handling	Wash thoroughly after handling. Remove contaminated clothing and wash before reuse. Use with adequate ventilation. Ground and bond containers when transferring material. Avoid contact with skin and eyes. Avoid ingestion and inhalation.

Hazards and Protection	
Protection	Wear appropriate eye protection and protective clothing to prevent skin and eye contact.
Respirators	Follow the OSHA respirator regulations found in 29CFR 1910.134 or European Standard EN 149. Always use a NIOSH or European Standard EN 149 approved respirator when necessary.
Small spills/leaks	Absorb spill with inert material, (e.g., dry sand or earth), then place into a chemical waste container. Remove all sources of ignition. Provide ventilation.
Stability	Stable at room temperature in closed containers under normal storage and handling conditions.
Incompatibilities	Chlorosulfonic acid, dimethyl terephthalate, oleum, phosphorus pentasulfide, silvered-copper wire, sodium hydroxide, sulfuric acid, titanium butoxide. Causes ignition at room temperature with chromium trioxide, potassium permanganate, and sodium peroxide. Causes ignition at 100 C with ammonium dichromate, silver chlorate, sodium chloride, and uranyl nitrate.
Decomposition	Carbon dioxide and carbon monoxide may form when heated to decomposition. May produce acrid smoke and irritating fumes when heated to decomposition.

Fire		
Flash Point,°C		111
Autoignition, °C		398
Upper exp. limit, %		15.3
Lower exp. limit, %		3.2
Fire fighting		Wear full protective clothing and NIOSH-approved self-contained breathing apparatus with full facepiece operated in the pressure demand or other positive pressure mode. Extinguish using dry chemical, foam or carbon dioxide. Water or foam may cause frothing. Water spray may be used to extinguish surrounding fire and cool exposed containers. Water spray will also reduce fume and irritant gases.
Fire potential		This chemical is combustible.
Combustion products		Toxic gases and vapors may be released if involved in a fire.
NFPA	**Health**	1
	Flammability	1
	Reactivity	0

Health	
Exposure limit(s)	TLV: 50 ppm; 127 mg/m^3 as CEILING (ACGIH 1993-1994). NIOSH REL: See Appendix D
Poison_Class	4
Exposure effects	Repeated small exposures by any route can cause severe kidney problems. Brain damage may also occur. Skin allergy can develop. May damage the developing fetus.

Health	
Ingestion	Initial symptoms in massive dosage parallel alcohol intoxication, progressing to CNS depression, vomiting, headache, rapid respiratory and heart rate, lowered blood pressure, stupor, collapse, and unconsciousness with convulsions. Death from respiratory arrest or cardiovascular collapse may follow. Lethal dose in humans: 100 ml.
Inhalation	Vapor inhalation is generally not a problem unless heated or misted. Exposure to vapors over an extended time period has caused throat irritation and headache. May cause nausea, vomiting, dizziness and drowsiness. Pulmonary edema and central nervous system depression may also develop. When heated or misted, has produced rapid, involuntary eye movement and coma.
Skin	May cause skin irritation. Low hazard for usual industrial handling.
Eyes	Splashes may cause irritation, pain, eye damage.

First aid	
Ingestion	If victim is conscious and alert, give 2-4 cupfuls of milk or water. Never give anything by mouth to an unconscious person. Get medical aid immediately.
Inhalation	Remove from exposure to fresh air immediately. If not breathing, give artificial respiration. If breathing is difficult, give oxygen. Get medical aid if cough or other symptoms appear.
Skin	Flush skin with plenty of soap and water for at least 15 minutes while removing contaminated clothing and shoes. Get medical aid if irritation develops or persists. Wash clothing before reuse.

(*Continued*)

First aid	
Eyes	First check the victim for contact lenses and remove if present. Flush victim's eyes with water or normal saline solution for 20 to 30 minutes while simultaneously calling a hospital or poison control center. Do not put any ointments, oils, or medication in the victim's eyes without specific instructions from a physician. IMMEDIATELY transport the victim after flushing eyes to a hospital even if no symptoms (such as redness or irritation) develop.

Transportation	
USCG CHRIS Code	EGL
USCG Compatatibility Group	20 Alcohols, Glycols
HS Code	2905 31 00
Std. Transport #	4960196
IMO Pollution Category	D

Formic Acid	
• Methanoic acid • Hydrogencarboxylic acid • Aminic acid • Formylic acid • Myrmicyl • Ameisensaure	
Formula	HCOOH
Structure	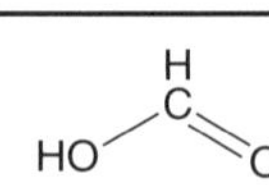
Description	Clear, colorless liquid that may fume, with an intense, pungent, penetrating odor.
Uses	Manufacture of fumigants.

Registry Numbers and Inventories	
CAS	64-18-6
NIH PubChem CID	284
EC (EINECS/ELINCS)	200-579-1
EC Index Number	607-001-00-0
EC Class	C; R35
RTECS	LQ4900000
RTECS class	Tumorigen; Mutagen; Human Data; Primary Irritant
UN (DOT)	1779
Merck	13,4265
Beilstein/Gmelin	1209246
Beilstein Reference	4-02-00-0003
RCRA	U123
EPA OPP	214900
FEMA	2487
Swiss Giftliste 1	G-4508
Canada DSL/NDSL	DSL
US TSCA	Listed
Austrailia AICS	Listed
New Zealand	Listed
Japan ENCS (MITI)	Listed
Korea ECL	Listed
Philippiens PICCS	Listed

Properties	
Formula	CH2O2
Formula mass	46.03
Melting point, °C	8
Boiling point, °C	100
Vapor pressure, mm_{Hg}	34 (20 C)
Vapor density (air=1)	1.6
Saturation Concentration	4.4% at 20 C
Evaporization number	2.1 (butyl acetate = 1)
Odor threshold	0.063 ppm
Critical temperature	315
Density	1.2183 g/cm^3 (20 C)
Solubility in water	Miscible
Viscosity	1.789 cp (20 C)
Surface tension	37.13 g/s2 @ 25 C
Refractive index	1.3714 (20 C)
Dipole moment	3.79 D
Dielectric constant	58.5 (16 C)
pKa/pKb	3.74 (pKa)
Partition coefficient, pK_{ow}	-0.54
Heat of fusion	12.7 kJ/mol
Heat of vaporization	20.1 kJ/mol
Heat of combustion	-255 kJ/mol

Hazards and Protection	
Storage	Store in a tightly closed container. Keep from contact with oxidizing materials. Store in a cool, dry, well-ventilated area away from incompatible substances. Corrosives area. Do not store in metal containers. Do not store near alkaline substances. Vent periodically
WHMIS	B3 E
Handling	Wash thoroughly after handling. Remove contaminated clothing and wash before reuse. Use only in a well ventilated area. Contents may develop pressure upon prolonged storage. Empty containers retain product residue, (liquid and/or vapor), and can be dangerous. Keep container tightly closed. Do not get on skin or in eyes. Do not ingest or inhale. Discard contaminated shoes. Do not pressurize, cut, weld, braze, solder, drill, grind, or expose empty containers to heat, sparks or open flames.
Protection	Rubber or neoprene gloves and additional protection including impervious boots, apron, or coveralls, as needed in areas of unusual exposure. Use chemical safety goggles and/or a full face shield where splashing is possible. Maintain eye wash fountain and quick-drench facilities in work area.
Respirators	Wear a supplied air, full-facepiece respirator, airlined hood, or full-facepiece self-contained breathing apparatus. Breathing air quality must meet the requirements of the OSHA respiratory protection standard (29CFR1910.134). Formic acid has questionable warning properties and a low IDLH. Respirator recommended to 6 times the TLV value as a maximum.

(*Continued*)

Hazards and Protection	
Small spills/leaks	Large spills may be neutralized with dilute alkaline solutions of soda ash, or lime. Avoid runoff into storm sewers and ditches which lead to waterways. Clean up spills immediately, using the appropriate protective equipment. Remove all sources of ignition. Absorb spill using an absorbent, non-combustible material such as earth, sand, or vermiculite. Use a spark-proof tool. Provide ventilation. Do not get water inside containers.
Stability	Stable under normal temperatures and pressures.
Incompatibilities	Sulfuric acid, strong caustics, furfuryl alcohol, hydrogen peroxide, strong oxidizers and bases. Reacts explosively with oxidizing agents.
Decomposition	Carbon monoxide, irritating and toxic fumes and gases, carbon dioxide.

Fire	
Flash Point,°C	50
Autoignition, °C	601
Upper exp. limit, %	57
Lower exp. limit, %	18

Fire		
Fire fighting		Wear a self-contained breathing apparatus in pressure-demand, MSHA/NIOSH (approved or equivalent), and full protective gear. During a fire, irritating and highly toxic gases may be generated by thermal decomposition or combustion. Use water spray to keep fire-exposed containers cool. Containers may explode in the heat of a fire. Extinguishing media: Do NOT get water inside containers. For small fires, use dry chemical, carbon dioxide, or water spray. For large fires, use dry chemical, carbon dioxide, alcohol-resistant foam, or water spray. Cool containers with flooding quantities of water until well after fire is out.
Fire potential		Moderately flammable. Combustion with moderate heating.
Hazards		When heated, vapors may form explosive mixtures with air: indoors, outdoors, and sewers explosion hazards. May polymerize explosively when involved in a fire. Contact with metals may evolve flammable hydrogen gas. Sensitive to static charge.
Combustion products		Toxic vapor generated in fires
NFPA	Health	3
	Flammability	2
	Reactivity	0

Health	
Exposure limit(s)	OSHA PEL: TWA 5 ppm (9 mg/m^3) NIOSH REL: TWA 5 ppm (9 mg/m^3) NIOSH IDLH: 30 ppm
Poison_Class	4
Exposure effects	Prolonged or repeated skin contact may cause dermatitis. May cause liver and kidney damage. Effects may be delayed. Laboratory experiments have resulted in mutagenic effects.
Ingestion	Causes severe digestive tract burns with abdominal pain, vomiting, and possible death. May be harmful if swallowed. May cause central nervous system depression. Ingestion may produce corrosive ulceration and bleeding and necrosis of the gastrointestinal tract accompanied by shock and circulatory collapse.
Inhalation	May cause asthmatic attacks due to allergic sensitization of the respiratory tract. Causes chemical burns to the respiratory tract. Aspiration may lead to pulmonary edema. Vapors may cause dizziness, nausea, itching, burning, and swelling of the eyes.
Skin	May cause skin sensitization, an allergic reaction, which becomes evident upon re-exposure to this material. Contact with liquid is corrosive and causes severe burns and ulceration. Absorbed through the skin. May cause erythema and blistering.
Eyes	Contact with liquid is corrosive to the eyes and causes severe burns. Lachrymator. May cause corneal edema, ulceration, and scarring.
First aid	

Health	
Ingestion	Do NOT induce vomiting. If victim is conscious and alert, give 2-4 cupfuls of milk or water. Never give anything by mouth to an unconscious person. Get medical aid immediately.
Inhalation	Get medical aid immediately. Remove from exposure to fresh air immediately. If breathing is difficult, give oxygen. DO NOT use mouth-to-mouth respiration. If breathing has ceased apply artificial respiration using oxygen and a suitable mechanical device such as a bag and a mask.
Skin	Get medical aid immediately. Immediately flush skin with plenty of soap and water for at least 15 minutes while removing contaminated clothing and shoes. Wash clothing before reuse. Destroy contaminated shoes.
Eyes	Get medical aid immediately. Do NOT allow victim to rub or keep eyes closed. Extensive irrigation is required (at least 30 minutes).

Transportation		
UN number	1779	
Response guide	153	
Hazard class	8	
Packing Group	II	
USCG CHRIS Code	FMA	
USCG Compatatibility Group	4 Organic acids	
HS Code	2915 11 00	

(Continued)

Transportation		
Std. Transport #	4931320	
IMO Chemical Code	17	
IMO Pollution Category	D	
IMO Hazard code	S	

Glutaraldehyde	
• 1,5-Pentanedial • Glutaric dialdehyde • Glutaral • 1,3-Diformylpropane	
Formula	$C_5H_8O_2$
Structure	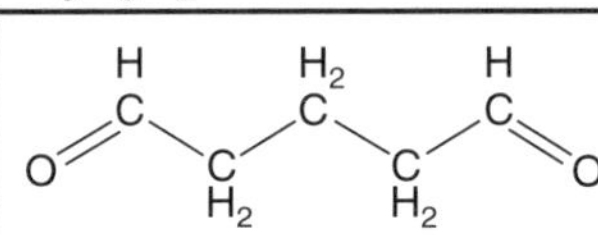
Description	Colorless to yellowish liquid with a pungent fruity/medicinal odor. A 5-25% aqueous solution. Physical properties are based on a 25% solution.
Uses	Microbiocide, fungicide.

Registry Numbers and Inventories	
CAS	111-30-8
NIH PubChem CID	3485
EC (EINECS/ELINCS)	203-856-5
EC Index Number	605-022-00-X
EC Class	T; R23/25, C; R34, R42/43, N; R50
RTECS	MA2450000
RTECS class	Tumorigen; Mutagen; Reproductive Effector; Primary Irritant

Registry Numbers and Inventories	
UN (DOT)	2810
Merck	13,4485
Beilstein/Gmelin	605390
Beilstein Reference	4-01-00-03659
EPA OPP	43901
Swiss Giftliste 1	G-42675
Canada DSL/NDSL	DSL
US TSCA	Listed
Austrailia AICS	Listed
New Zealand	Listed
Japan ENCS (MITI)	Listed
Korea ECL	Listed
Philippiens PICCS	Listed

Properties	
Formula	C5H8O2
Formula mass	100.12
Melting point, °C	-10
Boiling point, °C	101
Vapor pressure, mm_{Hg}	0.6 (25 C)
Vapor density (air=1)	0.8
Evaporization number	0.9 (butyl acetate=1)
Critical temperature	360
Critical pressure	41.1
Density	1.014 g/cm^3 (20 C)
Solubility in water	Miscible
Surface tension	<80 g/s2

(*Continued*)

Properties	
Refractive index	1.43 (20 C)
Partition coefficient, pK_{ow}	-0.34
Heat of vaporization	42.5 kJ/mol

Hazards and Protection	
Storage	Keep in a cool, dry, dark location in a tightly sealed container or cylinder. Keep away from incompatible materials, ignition sources and untrained individuals. Secure and label area. Protect containers/cylinders from physical damage.
WHMIS	D1B D2B E
Handling	All chemicals should be considered hazardous. Avoid direct physical contact. Use appropriate, approved safety equipment. Untrained individuals should not handle this chemical or its container. Handling should occur in a chemical fume hood.
Protection	Wear appropriate protective gloves, clothing and goggles.
Respirators	Wear positive pressure self-contained breathing apparatus (SCBA).
Small spills/leaks	ELIMINATE all ignition sources. Do not touch damaged containers or spilled material unless wearing appropriate protective clothing. Stop leak if you can do it without risk. Absorb or cover with dry earth, sand or other non-combustible material and transfer to containers. DO NOT GET WATER INSIDE CONTAINERS.

Hazards and Protection	
Stability	Stable under ordinary conditions of use and storage. May discolor on exposure to air. Polymerization may occur at high temperatures or in alkaline solutions (above pH 9), but it is not hazardous.
Incompatibilities	Avoid contact with strong oxidizers, strong bases.
Decomposition	When heated to decomposition it emits acrid smoke and fumes.

Fire		
Flash Point,°C		66
Fire fighting		Wear full protective clothing and NIOSH-approved self-contained breathing apparatus with full facepiece operated in the pressure demand or other positive pressure mode. Use any means suitable for extinguishing surrounding fire.
Fire potential		This material will not burn until the water has evaporated. Residue can burn.
Hazards		When heated, vapors may form explosive mixtures with air: indoors, outdoors, and sewers explosion hazards.
Combustion products		Fire may produce irritating, corrosive and/or toxic gases.
NFPA	**Health**	3
	Flammability	1
	Reactivity	0

Health	
Exposure limit(s)	TLV: 0.2 ppm; 0.82 mg/m^3 (ceiling values) (ACGIH 1992-1993). NIOSH REL: C 0.2 ppm (0.8 mg/m^3) See Appendix C (Aldehydes)
Poison_Class	3
Exposure effects	Seizures and CNS depression were noted in animals given intravenous glutaraldehyde.
Ingestion	Causes irritation to the gastrointestinal tract. Symptoms may include nausea, vomiting and diarrhea. May cause chemical burns of the mouth, throat, esophagus and stomach with discomfort or pain in the chest and abdomen. Aspiration into the lungs may occur during swallowing or vomiting, resulting in lung damage.
Inhalation	Causes irritation to the respiratory tract. Symptoms may include coughing, shortness of breath. Heating the solution may result in more severe irritant effects.
Skin	Causes irritation to skin. Symptoms include redness, itching, and pain. Prolonged contact may cause skin burns. May be absorbed through the skin with possible systemic effects.
Eyes	Corrosive! Vapors are irritating and may cause damage to the eyes. Contact may cause severe burns and permanent eye damage.
First aid	

Health	
Ingestion	Do not induce emesis - dilution: following ingestion and/or prior to gastric evacuation, immediately dilute with 4 to 8 ounces (120 to 240 ml) of milk or water (not to exceed 15 ml/kg in a child). The possible benefit of early removal of some ingested material by cautious gastric lavage must be weighed against potential complications of bleeding or perforation.
Inhalation	Move victim to fresh air. Apply artificial respiration if victim is not breathing. Do not use mouth-to-mouth method if victim ingested or inhaled the substance; induce artificial respiration with the aid of a pocket mask equipped with a one-way valve or other proper respiratory medical device. Administer oxygen if breathing is difficult.
Skin	Remove contaminated clothing and wash exposed area thoroughly with soap and water. A physician should examine the area if irritation or pain persists.
Eyes	Immediately flush with running water for at least 20 minutes.

Transportation		
UN number	2810	
Response guide	153	
Hazard class	6.1	
Packing Group	I; II; III	
USCG CHRIS Code	GTA	

(Continued)

Transportation		
USCG Compatatibility Group	19 Aldehydes	
IMO Chemical Code	17	
IMO Pollution Category	D	
IMO Hazard code	S	

Guar Gum	
• A-20d • 1212A • Burtonite v-7-e • Burtonite v 7e • Cyamopsis gum • Dealca tp1	
Formula	Unspecified
Description	Off-white to yellowish-white powder. Five to eight times the thickening power of starch.
Uses	As protective colloid, stabilizer, thickening and film forming agent for cheese, salad dressing, ice cream, soups, in paper sizing, as binding and disintegrating agent in tablet formulations, in pharmaceutical jelly formulations, in suspensions, emulsions, lotions, creams, toothpastes, in mining industry as flocculant, as filtering agent, in water treatment as a coagulant.

Registry Numbers and Inventories	
CAS	9000-30-0
NIH PubChem CID	47206285 (SID)
EC (EINECS/ELINCS)	232-536-8
RTECS	MG0185000
RTECS class	Tumorigen; Mutagen
Beilstein/Gmelin	NA
FEMA	2537
Canada DSL/NDSL	DSL
US TSCA	Listed
Austrailia AICS	Listed
New Zealand	Listed
Japan ENCS (MITI)	Listed
Korea ECL	Listed
Philippiens PICCS	Listed

Properties	
Solubility in water	completely soluble

Hazards and Protection	
Storage	Store in a tightly closed container. Store in a cool, dry, well-ventilated area away from incompatible substances.
Handling	Wash thoroughly after handling. Use with adequate ventilation. Minimize dust generation and accumulation. Avoid contact with eyes, skin, and clothing. Avoid ingestion and inhalation.

(Continued)

Hazards and Protection	
Protection	Eyes: Wear appropriate protective eyeglasses or chemical safety goggles as described by OSHA's eye and face protection regulations in 29 CFR 1910.133 or European Standard EN166. Skin: Wear appropriate protective gloves to prevent skin exposure. Clothing: Wear appropriate protective clothing to prevent skin exposure.
Respirators	Follow the OSHA respirator regulations found in 29CFR 1910.134 or European Standard EN 149. Always use a NIOSH or European Standard EN 149 approved respirator when necessary.
Small spills/leaks	Sweep up or absorb material, then place into a suitable clean, dry, closed container for disposal.
Stability	Stable under normal shipping and handling conditions.
Incompatibilities	Flammable and/or toxic gases are generated by combination with with alkali metals, nitrides, and strong reducing agents Reacts with oxoacids and carboxylic acids to form esters plus water.
Decomposition	Acrid smoke and fumes.

Fire	
Fire fighting	Wear a self-contained breathing apparatus in pressure-demand, MSHA/NIOSH (approved or equivalent), and full protective gear. Extinguishing media: Use agent most appropriate to extinguish fire.

Health	
Exposure effects	
Ingestion	May cause irritation of the digestive tract.
Inhalation	Inhalation of dust may cause respiratory tract irritation.
Skin	No information regarding skin irritation and other potential effects was found.
Eyes	Dust may cause mechanical irritation.
First aid	
Ingestion	If victim is conscious and alert, give 2-4 cupfuls of milk or water. Never give anything by mouth to an unconscious person. Get medical aid immediately.
Inhalation	Remove from exposure to fresh air immediately. If not breathing, give artificial respiration. If breathing is difficult, give oxygen. Get medical aid.
Skin	Get medical aid if irritation develops or persists.
Eyes	Flush eyes with plenty of water for at least 15 minutes, occasionally lifting the upper and lower eyelids. Get medical aid.

Hydrochloric Acid	
• Muriatic acid • Hydrogen chloride	
Formula	HCl
Structure	H—Cl
Description	An aqueous solution of hcl. Colorless, fuming liquid with a pungent odor.

(Continued)

Hydrochloric Acid	
Uses	In the production of chlorides, refining ore in the production of tin and tantalum, for the neutralization of basic systems, as laboratory reagent, hydrolyzing of starch and proteins in the preparation of various food products, pickling and cleaning of metal products, as catalyst and solvent in organic synthesis, for oil- and gas-well treatment, in removing scale from boilers and heat-exchange equipment, pharmaceutic aid (acidifier).

Registry Numbers and Inventories	
CAS	7647-01-0
NIH PubChem CID	313
EC (EINECS/ELINCS)	231-595-7
EC Index Number	017-002-01-X
EC Class	C; R34, Xi; R37
RTECS	MW4025000
RTECS class	Tumorigen; Mutagen; Reproductive Effector; Human Data; Primary Irritant
UN (DOT)	1789
Merck	13,4801
Beilstein/Gmelin	318 (G)
EPA OPP	45901
Swiss Giftliste 1	G-5658
Canada DSL/NDSL	DSL
US TSCA	Listed
Austrailia AICS	Listed

Registry Numbers and Inventories	
New Zealand	Listed
Japan ENCS (MITI)	Listed
Korea ECL	Listed
Philippiens PICCS	Listed
Israel	Listed

Properties	
Formula	ClH
Formula mass	36.46
Melting point, °C	-74
Boiling point, °C	109
Vapor pressure, mm_{Hg}	190 (25 C)
Vapor density (air=1)	1.25
Odor threshold	7.0 mg/m^3
Critical temperature	51.4
Critical pressure	81.6
Density	1.18 g/cm^3
Solubility in water	Miscible
Thermal expansion	0.003769/K
Heat of fusion	1.98 kJ/mol
Heat of vaporization	15.1 kJ/mol

Hazards and Protection	
Storage	Keep away from oxidizing agents, particularly nitric acid and chlorates. Safeguard containers against mechanical injury.
WHMIS	E D2A
Handling	Containers of this material may be hazardous when emptied. Since emptied containers retain product residues (vapor, liquid, and/or solid), all hazard precautions given in the data sheet must be observed. Emergency eyewash fountains and safety showers should be available in the immediate vicinity of potential exposure. Do not puncture or incinerate containers.
Protection	Wear appropriate chemical protective clothing.
Respirators	Wear positive pressure self-contained breathing apparatus.
Small spills/leaks	Keep material out of water sources and sewers. Attempt to stop leak if without undue personnel hazard. Use water spray to knock-down vapors. Neutralize spilled material with crushed limestone, soda ash, or lime. Do not use water on material itself. Vapor knockdown water is corrosive or toxic and should be diked for containment. Land spill: Dig a pit, pond, lagoon, holding area to contain liquid or solid material. Dike surface flow using soil, sand bags, foamed polyurethane, or foamed concrete.
Stability	Stable under ordinary conditions of use and storage. Containers may burst when heated.

Hazards and Protection	
Incompatibilities	Reacts rapidly and exothermically with bases of all kinds. Reacts exothermically with carbonates (and hydrogen carbonates to generate carbon dioxide Reacts with sulfides, carbides, borides, phosphides, many metals (including aluminum, zinc, calcium, magnesium, iron, tin and all of the alkali metals) to generate flammable hydrogen gas Reacts violently with acetic anhydride, 2-aminoethanol, ammonium hydroxide, calcium phosphide, chlorosulfonic acid, 1,1-difluoroethylene, ethylenediamine, ethyleneimine, oleum, perchloric acid, b-propiolactone, propylene oxide, silver perchlorate/carbon tetrachloride mixture, sodium hydroxide, uranium(IV) phosphide, vinyl acetate, calcium carbide, rubidium carbide, cesium acetylide, rubidium acetylide, magnesium boride, mercury(II) sulfate.
Decomposition	When heated to decomposition, emits toxic hydrogen chloride fumes and will react with water or steam to produce heat and toxic and corrosive fumes. Thermal oxidative decomposition produces toxic chlorine fumes and explosive hydrogen gas.

Fire		
Fire fighting		Wear full protective clothing and NIOSH-approved self-contained breathing apparatus with full facepiece operated in the pressure demand or other positive pressure mode. Structural firefighter's protective clothing is ineffective for fires involving hydrochloric acid. Stay away from ends of tanks. Extinguish fire using agent suitable for type of surrounding fire. (Material itself does not burn.) Cool all affected containers with flooding quantities of water. Apply water from as far a distance as possible. Use water spray to knock-down vapors.
Fire potential		Extreme heat or contact with metals can release flammable hydrogen gas
Hazards		Vapors from liquefied gas are initially heavier than air and spread along ground. Containers may explode when heated. Ruptured cylinders may rocket.
Combustion products		Non-combustible, substance itself does not burn but may decompose upon heating to produce corrosive and/or toxic fumes.
NFPA	**Health**	3
	Flammability	0
	Reactivity	1

Health	
Exposure limit(s)	TLV: 5 ppm; 7.5 mg/m^3 (ceiling values) (ACGIH 1992-1993). OSHA PEL: C 5 ppm (7 mg/m^3) NIOSH REL: C 5 ppm (7 mg/m^3) NIOSH IDLH: 50 ppm
Poison_Class	4
Exposure effects	Long-term exposure to concentrated vapors may cause erosion of teeth. Long term exposures seldom occur due to the corrosive properties of the acid.
Ingestion	Corrosive! Swallowing hydrochloric acid can cause immediate pain and burns of the mouth, throat, esophagus and gastrointestinal tract. May cause nausea, vomiting, and diarrhea. Swallowing may be fatal.
Inhalation	Corrosive! Inhalation of vapors can cause coughing, choking, inflammation of the nose, throat, and upper respiratory tract, and in severe cases, pulmonary edema, circulatory failure, and death.
Skin	Corrosive! Can cause redness, pain, and severe skin burns. Concentrated solutions cause deep ulcers and discolor skin.
Eyes	Corrosive! Vapors are irritating and may cause damage to the eyes. Contact may cause severe burns and permanent eye damage.
First aid	
Ingestion	Have person drink water or milk; do NOT induce vomiting.
Inhalation	Remove person to fresh air; keep him warm and quiet and get medical attention immediately; start artificial respiration if breathing stops.

(*Continued*)

First aid	
Skin	Remove and isolate contaminated clothing and shoes. Immediately flush with running water for at least 20 minutes. For minor skin contact, avoid spreading material on unaffected skin.
Eyes	Irrigate exposed eyes with copious amounts of tepid water for at least 15 minutes. If irritation, pain, swelling, lacrimation, or photophobia persist, the patient should be seen in a health care facility.

Transportation		
UN number	1789	
Response guide	157	
Hazard class	8	
Packing Group	II; III	
USCG CHRIS Code	HDC	
USCG Compatatibility Group	1 Non-oxidizing mineral acids	
Std. Transport #	4920504	
IMO Chemical Code	17	
IMO Pollution Category	D	
IMO Hazard code	S	

Isopropanol	
• 2-Propanol • Isopropyl alcohol	
Propan-2-ol	
Formula	$CH_3CH(OH)CH_3$
Structure	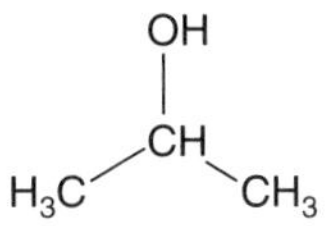
Description	Colorless liquid with a sharp, musty odor of rubbing alcohol.
Uses	Microbiocide, solvent.

Registry Numbers and Inventories	
CAS	67-63-0
NIH PubChem CID	3776
EC (EINECS/ELINCS)	200-661-7
EC Index Number	603-117-00-0
EC Class	F; R11, Xi; R36, R67
RTECS	NT8050000
RTECS class	Agricultural Chemical and Pesticide; Tumorigen; Drug; Mutagen; Reproductive Effector; Human Data; Primary Irritant
UN (DOT)	1219
Merck	13,5228
Beilstein/Gmelin	635639
Beilstein Reference	4-01-00-01461
EPA OPP	47501
FEMA	2929

Registry Numbers and Inventories	
Swiss Giftliste 1	G-1712
Canada DSL/NDSL	DSL
US TSCA	Listed
Austrailia AICS	Listed
New Zealand	Listed
Japan ENCS (MITI)	Listed
Korea ECL	Listed
Philippiens PICCS	Listed

Properties	
Formula	C3H8O
Formula mass	60.10
Melting point, °C	-89
Boiling point, °C	82
Vapor pressure, mm_{Hg}	44 (25 C)
Vapor density (air=1)	2.1
Saturation Concentration	43600 ppm (4.36%) at 20 C; 59700 ppm (5.97%) at 25 C (calculated)
Evaporization number	2.83 (butyl acetate=1)
Odor threshold	90 mg/m^3
Critical temperature	235
Critical pressure	47.00
Density	0.786 g/cm^3
Solubility in water	Miscible
Viscosity	2.08 cp (25 C)
Surface tension	21.19 g/s2 (25 C)

Properties	
Refractive index	0.3852 (20 C)
Dipole moment	1.66 D (20 C)
Dielectric constant	18.3 (25 C)
Partition coefficient, pK_{ow}	0.05
Heat of fusion	5.4 kJ/mol
Heat of vaporization	28.3 kJ/mol
Heat of combustion	-1995 kJ/mol

Hazards and Protection	
Storage	Keep away from sources of ignition. Store in a cool, dry place. Store in a tightly closed container.
WHMIS	B2 D2A D2B
Handling	Wash thoroughly after handling. Wash hands before eating. Use only in a well ventilated area. Use spark-proof tools and explosion proof equipment. Empty containers retain product residue, (liquid and/or vapor), and can be dangerous. Do not get on skin or in eyes. Avoid ingestion and inhalation. Do not pressurize, cut, weld, braze, solder, drill, grind, or expose empty containers to heat, sparks or open flames.
Protection	Wear impervious protective clothing, including boots, gloves, lab coat, apron or coveralls, as appropriate, to prevent skin contact. Neoprene and nitrile rubber are recommended materials. Use chemical safety goggles and/or a full face shield where splashing is possible. Maintain eye wash fountain and quick-drench facilities in work area.

(*Continued*)

Hazards and Protection	
Respirators	A full facepiece respirator with organic vapor cartridge may be worn up to 50 times the exposure limit or the maximum use concentration specified by the appropriate regulatory agency or respirator supplier, whichever is lowest. For emergencies or instances where the exposure levels are not known, use a full-facepiece positive-pressure, air-supplied respirator.
Small spills/leaks	Absorb spill with inert material, (e.g., dry sand or earth), then place into a chemical waste container. Clean up spills immediately, using the appropriate protective equipment. Scoop up with a nonsparking tool, then place into a suitable container for disposal. Remove all sources of ignition.
Disposal code	1
Stability	Stable at room temperature in closed containers under normal storage and handling conditions. This material may be sensitive to peroxide formation. Heat and sunlight can contribute to instability.
Incompatibilities	Oxidizing agents, aldehydes, acetaldeyde, chlorine, ethylene oxide, acids and isocyanates, hydrogen + palladium, nitroform, oleum, phosgene, potassium t-butoxide, oxygen, trinitromethane, barium perchlorate, tetrafluoroborate, nitric acid, halogen and halogen compounds, chromium trioxide, sodium dichromate + sulfuric acid, aluminum, and aluminum triisopropoxide. It has also been reported to be susceptible to autoxidation and should therRefer to therefore be considered peroxidizable. NFPA manual for more specific information.
Decomposition	Carbon monoxide, carbon dioxide.

Fire		
Flash Point,°C		12
Autoignition, °C		399
Upper exp. limit, %		12.7
Lower exp. limit, %		2.0
Fire fighting		Use water in flooding quantities as fog. Use foam, dry chemical, or carbon dioxide. Cool all affected containers with flooding quantities of water. Apply water from as far a distance as possible. Use water spray to knock-down vapors. Keep run-off water out of sewers and water sources.
Fire potential		HIGHLY FLAMMABLE.
Hazards		Vapors may form explosive mixtures with air. Vapors may travel to source of ignition and flash back. Most vapors are heavier than air. They will spread along ground and collect in low or confined areas (sewers, basements, tanks). Vapor explosion hazard indoors, outdoors or in sewers. May polymerize explosively when heated or involved in a fire. Runoff to sewer may create fire or explosion hazard. Containers may explode when heated. Many liquids are lighter than water.
Combustion products		Fire may produce irritating, corrosive and/or toxic gases.
NFPA	Health	1
	Flammability	3
	Reactivity	0

Health	
Exposure limit(s)	OSHA PEL: TWA 400 ppm (980 mg/m^3) NIOSH REL: TWA 400 ppm (980 mg/m^3) ST 500 ppm (1225 mg/m^3) NIOSH IDLH: 2000 ppm LEL
Poison_Class	-
Exposure effects	Prolonged or repeated skin contact may cause defatting and dermatitis. May cause allergic skin reaction in some individuals.
Ingestion	Can cause drowsiness, unconsciousness, and death. Gastrointestinal pain, cramps, nausea, vomiting, and diarrhea may also result. The single lethal dose for a human adult = about 250 mls (8 ounces).
Inhalation	Inhalation of high concentrations may cause central nervous system effects characterized by headache, dizziness, unconsciousness and coma. Inhalation of vapor may cause respiratory tract irritation. May cause narcotic effects.
Skin	May cause skin sensitization, an allergic reaction, which becomes evident upon re-exposure to this material. Prolonged and/or repeated contact may cause defatting of the skin and dermatitis. May cause irritation with pain and stinging, especially if the skin is abraded.
Eyes	Produces irritation, characterized by a burning sensation, redness, tearing, inflammation, and possible corneal injury.

First aid	
Ingestion	If victim is conscious and alert, give 2-4 cupfuls of milk or water. Never give anything by mouth to an unconscious person. Get medical aid immediately. Induce vomiting by giving one teaspoon of Syrup of Ipecac.

Health	
Inhalation	Get medical aid immediately. Remove from exposure to fresh air immediately. If not breathing, give artificial respiration. If breathing is difficult, give oxygen.
Skin	Get medical aid if irritation develops or persists. Flush skin with plenty of soap and water.
Eyes	First check the victim for contact lenses and remove if present. Flush victim's eyes with water or normal saline solution for 20 to 30 minutes while simultaneously calling a hospital or poison control center. Do not put any ointments, oils, or medication in the victim's eyes without specific instructions from a physician. IMMEDIATELY transport the victim after flushing eyes to a hospital even if no symptoms (such as redness or irritation) develop.

Transportation		
UN number	1219	
Response guide	129	
Hazard class	3	
Packing Group	II	
USCG CHRIS Code	IPA	
USCG Compatatibility Group	20 Alcohols, Glycols	
HS Code	2905 12 00	
Std. Transport #	4907810	
IMO Chemical Code	18	
IMO Pollution Category	III	

Magnesium Oxide	
• Magnesia • Calcined brusite • Calcined magnesia	
Formula	MgO
Structure	Mg^{2+} O^{2-}
Description	Fine odorless white powder, hygroscopic.
Uses	Chemical intermediate medicinals & other chems, component of animal feeds, fertilizers, insulation, & wallboard, component of petroleum additives, oxychloride & oxysulfate cement, agent in pulp & paper mfr, food chem, rubber accelerator, component of electrical heating rods, misc applications.

Registry Numbers and Inventories	
CAS	1309-48-4
NIH PubChem CID	14792
EC (EINECS/ELINCS)	215-171-9
EC Class	S: 22
RTECS	OM3850000
RTECS class	Tumorigen; Human Data
UN (DOT)	1418
Merck	12,5713
Beilstein/Gmelin	13722 (G)
EPA OPP	9235
Swiss Giftliste 1	G-2368
Canada DSL/NDSL	DSL

Registry Numbers and Inventories	
US TSCA	Listed
Austrailia AICS	Listed
New Zealand	Listed
Japan ENCS (MITI)	Listed
Korea ECL	Listed
Philippiens PICCS	Listed

Properties	
Formula	MgO
Formula mass	40.30
Melting point, °C	2800
Boiling point, °C	3600
Vapor pressure, mm_{Hg}	0.3 (1727 C)
Density	3.58 g/cm^3
Solubility in water	90 g/L
Dielectric constant	11 (20 C)
Thermal expansion	1.38E-5/L
Heat of fusion	77.4 kJ/mol
Heat of vaporization	330.7 kJ/mol

Hazards and Protection	
Storage	Store in a cool, dry place. Keep away from water. Keep containers tightly closed.
WHMIS	Does not meet criteria

(*Continued*)

Hazards and Protection	
Handling	Wash thoroughly after handling. Avoid contact with eyes, skin, and clothing. Keep container tightly closed. Avoid ingestion and inhalation. Use with adequate ventilation. Wash clothing before reuse. Keep from contact with moist air and steam.
Protection	Eyes: Wear appropriate protective eyeglasses or chemical safety goggles as described by OSHA's eye and face protection regulations in 29 CFR 1910.133 or European Standard EN166. Skin: Wear appropriate protective gloves to prevent skin exposure. Clothing: Wear appropriate protective clothing to prevent skin exposure.
Respirators	Follow the OSHA respirator regulations found in 29CFR 1910.134 or European Standard EN 149. Always use a NIOSH or European Standard EN 149 approved respirator when necessary.
Small spills/leaks	Wash area with soap and water. Clean up spills immediately, using the appropriate protective equipment. Sweep up, then place into a suitable container for disposal. Avoid generating dusty conditions. Do not get water inside containers.
Stability	Stable under normal temperatures and pressures.
Incompatibilities	Strong oxidizing agents; reacts violently with phosphorous pentachloride, chlorine trichloride, or bromine pentafluoride. Will absorb CO2 from air.
Decomposition	None.

Fire		
Fire fighting		Wear a self-contained breathing apparatus in pressure-demand, MSHA/NIOSH (approved or equivalent), and full protective gear. During a fire, irritating and highly toxic gases may be generated by thermal decomposition or combustion. To extinguish fire use water spray, dry chemical, carbon dioxide, or chemical foam. Do NOT get water inside containers.
NFPA	**Health**	1
	Flammability	0
	Reactivity	0

Health	
Exposure limit(s)	OSHA PEL: TWA 15 mg/m^3 NIOSH REL: See Appendix D NIOSH IDLH: 750 mg/m^3
Poison_Class	4
Exposure effects	Lethargy, hyporeflexia, weakness, and subjective feeling of thirst and heat may occur. Excessive absorption of magnesium from cathartics may result in CNS depression and seizures, most notably in renal failure patients.
Ingestion	May cause gastrointestinal irritation with nausea, vomiting and diarrhea. May cause central nervous system depression.

(Continued)

Health	
Inhalation	May cause respiratory tract irritation. Inhalation of fumes may cause metal fume fever, which is characterized by flu-like symptoms with metallic taste, fever, chills, cough, weakness, chest pain, muscle pain and increased white blood cell count.
Skin	May cause skin irritation.
Eyes	May cause eye irritation.
First aid	
Ingestion	If victim is conscious and alert, give 2-4 cupfuls of milk or water. Never give anything by mouth to an unconscious person. Get medical aid immediately.
Inhalation	Remove from exposure to fresh air immediately. If not breathing, give artificial respiration. If breathing is difficult, give oxygen. Get medical aid.
Skin	Get medical aid if irritation develops or persists. Wash clothing before reuse. Flush skin with plenty of soap and water.
Eyes	Immediately flush eyes with plenty of water for at least 15 minutes, occasionally lifting the upper and lower eyelids. Get medical aid immediately.

Transportation	
UN number	1418
Response guide	138
USCG CHRIS Code	MGU
HS Code	2519 90 10

Magnesium Peroxide	
• Magnesium dioxide	
Formula	MgO_2
Structure	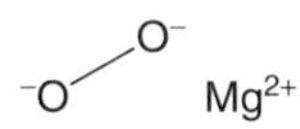
Description	a white powder.
Uses	Used in medicine, and as a bleaching agent.

Registry Numbers and Inventories	
CAS	1335-26-8
NIH PubChem CID	159386
EC (EINECS/ELINCS)	215-627-7
UN (DOT)	1483
Merck	12,5717
Beilstein/Gmelin	95156 (G)
Canada DSL/NDSL	NDSL
US TSCA	Listed
New Zealand	Listed
Korea ECL	Listed

Properties	
Formula	MgO2
Formula mass	56.30
Melting point, °C	222 - 225
Density	~3.0 g/cm^3 (20 C)
Solubility in water	Insoluble

Hazards and Protection	
Storage	Store in a cool, dry location in a tightly sealed container.
Handling	All chemicals should be considered hazardous. Avoid direct physical contact. Use appropriate, approved safety equipment. Untrained individuals should not handle this chemical or its container. Handling should occur in a chemical fume hood.
Protection	Wear protective gloves and clothing and safety goggles.
Respirators	Follow the OSHA respirator regulations found in 29CFR 1910.134 or European Standard EN 149. Always use a NIOSH or European Standard EN 149 approved respirator when necessary.
Small spills/leaks	Keep sparks, flames, and other sources of ignition away. Keep material out of water sources and sewers.
Stability	No data.

Fire	
Fire fighting	Flood with water. Cool all affected containers with flooding quantities of water. Apply water from as far a distance as possible.
Fire potential	Oxidizing material. Contact with combustible materials may cause fire.
Hazards	These substances will accelerate burning when involved in a fire. Some may decompose explosively when heated or involved in a fire. May explode from heat or contamination. Some will react explosively with hydrocarbons (fuels). May ignite combustibles (wood, paper, oil, clothing, etc.). Containers may explode when heated. Runoff may create fire or explosion hazard.

Fire	
Combustion products	Fire may produce irritating, corrosive and/or toxic gases.
Special	O

Health	
Exposure effects	
Ingestion	May causes irritation. May be harmful if ingested.
Inhalation	May causes irritation. May be harmful if inhaled.
Skin	May causes irritation. May be harmful if absorbed.
Eyes	May cause irritation.
First aid	
Ingestion	Wash mouth out with water, provided that the person is concious. Seek medical attention.
Inhalation	Remove to fresh air. If no breathing, give artificial respiration. If breathing is difficult, give oxygen.
Skin	Immediately wash skin with soap and large amounts of water.
Eyes	Flush eyes with copious amounts of water for at least 15 minutes. Seek medical attention.

Transportation		
UN number	1483	
Response guide	140	
Hazard class	5.1	
Packing Group	II; III	
Std. Transport #	4918730	

Methanol	
• (2H4)Methanol • Tetradeuteromethanol • Perdeuteromethanol	
Formula	CD_3OD
Structure	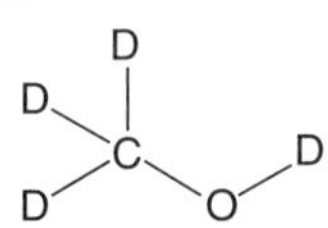
Description	Colorless liquid. Alcohol-like odor.

Registry Numbers and Inventories	
CAS	811-98-3
NIH PubChem CID	71568
EC (EINECS/ELINCS)	212-378-6
EC Class	F; T, R: 11-23/24/25-39/23/24/25, S: 7-16-36/37-45
UN (DOT)	1230
Beilstein/Gmelin	NA
Canada DSL/NDSL	NDSL
US TSCA	Listed
Austrailia AICS	Listed
New Zealand	Listed
Philippiens PICCS	Listed

Properties	
Formula	CH4O
Formula mass	36.07
Melting point, °C	-99
Boiling point, °C	65.3
Density	0.888 g/cm^3
Solubility in water	Soluble

Hazards and Protection	
Storage	Keep away from heat, sparks, and flame. Keep away from sources of ignition. Store in a tightly closed container. Store in a cool, dry, well-ventilated area away from incompatible substances.
Handling	Wash thoroughly after handling. Use with adequate ventilation. Ground and bond containers when transferring material. Use spark-proof tools and explosion proof equipment. Avoid contact with eyes, skin, and clothing. Empty containers retain product residue, (liquid and/or vapor), and can be dangerous. Keep container tightly closed. Avoid contact with heat, sparks and flame. Do not ingest or inhale. Do not pressurize, cut, weld, braze, solder, drill, grind, or expose empty containers to heat, sparks or open flames.
Protection	Wear appropriate protective gloves, clothing and goggles.

(*Continued*)

Hazards and Protection	
Respirators	Follow the OSHA respirator regulations found in 29CFR 1910.134 or European Standard EN 149. Always use a NIOSH or European Standard EN 149 approved respirator when necessary.
Small spills/leaks	Absorb spill with inert material, (e.g., dry sand or earth), then place into a chemical waste container. Remove all sources of ignition. A vapor suppressing foam may be used to reduce vapors. Water spray may reduce vapor but may not prevent ignition in closed spaces.
Disposal code	1
Stability	Stable under normal temperatures and pressures.
Incompatibilities	Acids, methanoic acid, oxalic acid), azo, diazo, and hydrazines (e.g. dimethyl hydrazine, hydrazine, methyl hydrazine), isocyanates (e.g. methyl isocyanate), metals (alkali and alkaline, e.g. cesium, potassium, sodium), nitrides (e.g. potassium nitride, sodium nitride), peroxides and hydroperoxides (organic, e.g. acetyl peroxide, benzoyl peroxide, butyl peroxide, methyl ethyl ketone peroxide), epoxides (e.g. butyl glycidyl ether), oxidizing agents (strong, e.g. bromine, hydrogen peroxide, nitrogen dioxide, potassium nitrate), reducing agents (strong, e.g. aluminum carbide, chlorosilane, hydrogen phosphide, lithium hydride), water reactive substances (e.g. acetic anyhdride, alkyl aluminum chlori
Decomposition	Carbon monoxide, carbon dioxide, formaldehyde.

Fire	
Flash Point,°C	12
Autoignition, °C	455
Upper exp. limit, %	31
Lower exp. limit, %	5.5
Fire fighting	Use of water spray when fighting fire may be inefficient. Small Fires: Dry chemical, carbon dioxide, water spray or alcohol-resistant foam. Large Fires: Water spray, fog or alcohol-resistant foam.
Fire potential	HIGHLY FLAMMABLE: Will be easily ignited by heat, sparks or flames.
Hazards	Vapors may form explosive mixtures with air. Vapors may travel to source of ignition and flash back. Most vapors are heavier than air. They will spread along ground and collect in low or confined areas (sewers, basements, tanks). Vapor explosion and poison hazard indoors, outdoors or in sewers.
Combustion products	Fire will produce irritating, corrosive and/or toxic gases.

Health	
Poison_Class	3 (Strong toxins)
Exposure effects	Prolonged or repeated skin contact may cause dermatitis. Chronic inhalation and ingestion may cause effects similar to those of acute inhalation and ingestion.

Ingestion	May be fatal or cause blindness if swallowed. May cause kidney damage. May cause systemic toxicity with acidosis. May cause central nervous system depression, characterized by excitement, followed by headache, dizziness, drowsiness, and nausea. Advanced stages may cause collapse, unconsciousness, coma and possible death due to respiratory failure.
Inhalation	May cause respiratory tract irritation. May cause visual impairment and possible permanent blindness. May cause effects similar to those described for ingestion. May cause drowsiness, unconsciousness, and central nervous system depression.
Skin	May cause skin irritation. May be absorbed through the skin.
Eyes	Causes eye irritation. Vapors may cause eye irritation. May cause painful sensitization to light.
First aid	
Ingestion	If victim is conscious and alert, give 2-4 cupfuls of milk or water. Never give anything by mouth to an unconscious person. Get medical aid immediately. Induce vomiting by giving one teaspoon of Syrup of Ipecac.
Inhalation	Remove from exposure to fresh air immediately. If not breathing, give artificial respiration. If breathing is difficult, give oxygen. Get medical aid.
Skin	Flush skin with plenty of soap and water for at least 15 minutes while removing contaminated clothing and shoes. Get medical aid if irritation develops or persists.
Eyes	In case of contact with substance, immediately flush with running water for at least 20 minutes.

Transportation		
UN number	1230	
Response guide	131	
Hazard class	3	
Packing Group	II	
HS Code	2845 90 10	

Naphthalene	
• Naphthalene, crude • Naphthalene, refined • Tar camphor • White tar • Mothballs • Naphthaliin	
Formula	$C_{10}H_8$
Structure	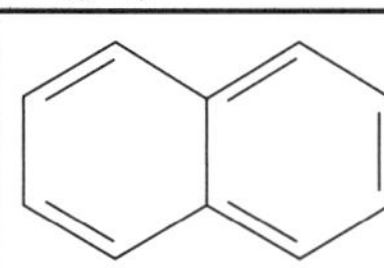
Description	White crystalline solid with distinctive odor of mothballs or coal tar.
Uses	Manufacture of phthalic & anthranilic acids, naphthols, naphthylamines, sulfonic acid, synthetic resins, celluloid, lampblack, smokeless powder, hydronaphthalenes, insecticide.

Registry Numbers and Inventories	
CAS	91-20-3
NIH PubChem CID	931
EC (EINECS/ELINCS)	202-049-5
EC Index Number	601-052-00-2
EC Class	Carc. Cat. 3; R40, Xn; R22, N; R50-53
RTECS	QJ0525000
RTECS class	Agricultural Chemical and Pesticide; Tumorigen; Mutagen; Reproductive Effector; Human Data; Primary Irritant
UN (DOT)	1334
Merck	12,6457
Beilstein/Gmelin	1421310
Beilstein Reference	4-05-00-01640
RCRA	U165
EPA OPP	55801
Swiss Giftliste 1	G-2069
Canada DSL/NDSL	DSL
US TSCA	Listed
Austrailia AICS	Listed
New Zealand	Listed
Japan ENCS (MITI)	Listed
Korea ECL	Listed
Philippiens PICCS	Listed

Properties	
Formula	C10H8
Formula mass	128.16

Properties	
Melting point, °C	80
Boiling point, °C	218
Vapor pressure, mm_{Hg}	0.08 (25 C)
Vapor density (air=1)	4.42
Saturation Concentration	100 ppm at 25 C
Evaporization number	<1 (butyl acetate=1)
Odor threshold	0.084 ppm
Critical temperature	475
Critical pressure	40.0
Density	0.979 g/cm^3 (80 C)
Solubility in water	30 mg/L (20 C)
Viscosity	0.886 cp (80 C)
Surface tension	31.8 g/s2
Refractive index	1.58212 (20 C)
Dipole moment	0.92 D
Dielectric constant	3.22 (81 C)
Partition coefficient, pK_{ow}	3.30
Heat of fusion	18.8 kJ/mol
Heat of vaporization	43.5 kJ/mol
Heat of combustion	-5196

Hazards and Protection	
Storage	Keep away from sources of ignition. Store in a tightly closed container. Store in a cool, dry, well-ventilated area away from incompatible substances. Store protected from moisture.

(Continued)

Hazards and Protection	
WHMIS	B4 D2B
Handling	Wash thoroughly after handling. Remove contaminated clothing and wash before reuse. Minimize dust generation and accumulation. Avoid contact with eyes, skin, and clothing. Do not breathe dust, vapor, mist, or gas. Keep container tightly closed. Do not ingest or inhale. Use only in a chemical fume hood.
Protection	Eyes: Wear appropriate protective eyeglasses or chemical safety goggles as described by OSHA's eye and face protection regulations in 29 CFR 1910.133 or European Standard EN166. Skin: Wear appropriate protective gloves to prevent skin exposure. Clothing: Wear appropriate protective clothing to prevent skin exposure.
Respirators	A half-face respirator with an organic vapor cartridge and particulate filter (NIOSH type P95 or R95 filter) may be worn for up to ten times the exposure limit or the maximum use concentration specified by the appropriate regulatory agency or respirator supplier, whichever is lowest. A full-face piece respirator with an organic vapor cartridge and particulate filter (NIOSH P100 or R100 filter) may be worn up to 50 times the exposure limit, or the maximum use concentration specified by the appropriate regulatory agency or respirator supplier, whichever is lowest. Please note that N series filters are not recommended for this material. For emergencies or instances where the exposure levels are not known, use a full-face piece positive-pressure, air-supplied respirator.

Hazards and Protection	
Small spills/leaks	Avoid runoff into storm sewers and ditches which lead to waterways. Clean up spills immediately, using the appropriate protective equipment. Scoop up with a nonsparking tool, then place into a suitable container for disposal. Avoid generating dusty conditions. Remove all sources of ignition.
Disposal code	3
Stability	Stable at room temperature in sealed containers. Sublimes appreciably at temperatures above melting point.
Incompatibilities	Strong oxidizers, strong alkalis and strong mineral acids, mixtures of aluminum trichloride and benzoyl chloride. Reacts violently with chromic anhydride.
Decomposition	Carbon monoxide, irritating and toxic fumes and gases, carbon dioxide.

Fire	
Flash Point,°C	87
Autoignition, °C	526
Upper exp. limit, %	5.9
Lower exp. limit, %	0.9

(*Continued*)

Fire		
Fire fighting		Wear a self-contained breathing apparatus in pressure-demand, MSHA/NIOSH (approved or equivalent), and full protective gear. Water runoff can cause environmental damage. Dike and collect water used to fight fire. Use water spray to keep fire-exposed containers cool. Dust can be an explosion hazard when exposed to heat or flame. May be ignited by friction, heat, sparks, or flame. May re-ignite after fire is extinguished. Containers may explode when heated. Extinguishing media: Use dry sand or earth to smother fire. Water or foam may cause frothing. Cool containers with flooding quantities of water until well after fire is out. Use dry chemical, carbon dioxide, or appropriate foam.
Fire potential		Flammable/combustible material.
Hazards		May be ignited by friction, heat, sparks or flames. Some may burn rapidly with flare burning effect. Powders, dusts, shavings, borings, turnings or cuttings may explode or burn with explosive violence. Substance may be transported in a molten form. May re-ignite after fire is extinguished.
Combustion products		Toxic vapors given off in a fire.
NFPA	**Health**	2
	Flammability	2
	Reactivity	0

Health	
Exposure limit(s)	NIOSH REL: TWA 10 ppm (50 mg/m^3) ST 15 ppm (75 mg/m^3) OSHA PEL: TWA 10 ppm (50 mg/m^3) IDLH 250 ppm
Poison_Class	3
Exposure effects	Prolonged or repeated skin contact may cause dermatitis. May cause liver and kidney damage. May cause anemia and other blood cell abnormalities. Animal studies have reported that fetal effects/abnormalities may occur when maternal toxicity is seen. Effects may be delayed. Chronic exposure may cause lung damage. Laboratory experiments have resulted in mutagenic effects. Chronic exposure may cause corneal injury, optical neuritis, blurred vision, and possible cataract formation.
Ingestion	Toxic. Can cause headache, profuse perspiration, listlessness, dark urine, nausea, vomiting and disorientation. Intravascular hemolysis may also occur with symptoms similar to those noted for inhalation. Severe cases may produce coma with or without convulsions. Death may result from renal failure
Inhalation	Inhalation of dust or vapors can cause headache, nausea, vomiting, extensive sweating, and disorientation. The predominant reaction is delayed intravascular hemolysis with symptoms of anemia, fever, jaundice, and kidney or liver damage.
Skin	Causes mild skin irritation. May be absorbed through the skin in harmful amounts. May cause skin sensitization, an allergic reaction, which becomes evident upon re-exposure to this material.

(*Continued*)

Health	
Eyes	Vapors and solid causes irritation, redness and pain. Very high exposures can damage the nerves of the eye.

First aid	
Ingestion	Do NOT induce vomiting. If victim is conscious and alert, give 2-4 cupfuls of milk or water. Never give anything by mouth to an unconscious person. Get medical aid immediately.
Inhalation	Remove from exposure to fresh air immediately. If breathing is difficult, give oxygen. Get medical aid. DO NOT use mouth-to-mouth respiration. If breathing has ceased apply artificial respiration using oxygen and a suitable mechanical device such as a bag and a mask.
Skin	Get medical aid. Immediately flush skin with plenty of soap and water for at least 15 minutes while removing contaminated clothing and shoes. Wash clothing before reuse.
Eyes	Immediately flush eyes with plenty of water for at least 15 minutes, occasionally lifting the upper and lower eyelids. Get medical aid.

Transportation		
UN number	1334	
Response guide	133	
Hazard class	4.1	
Packing Group	III	
USCG CHRIS Code	NTM	

Transportation		
USCG Compatatibility Group	32 Aromatic hydrocarbons	
HS Code	2902 90 10	
Std. Transport #	4917459 4917473	
IMO Chemical Code	17	
IMO Pollution Category	A	
IMO Hazard code	S/P	

Potassium Carbonate	
• Carbonate of potash • Dipotassium carbonate • Potash • Pearl ash	
Formula	K_2CO_3
Structure	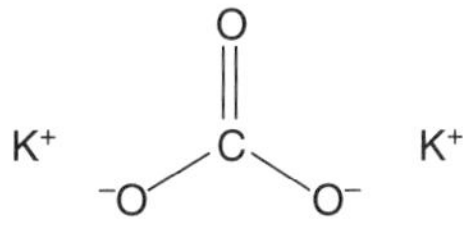
Description	White granules or granular, translucent powder, small glassy colorless crystals, colorless monoclinic crystals. Hygroscopic.
Uses	Ph adjustment, fungicide, microbiocide, herbicide.

Registry Numbers and Inventories	
CAS	584-08-7
NIH PubChem CID	11430

(Continued)

Registry Numbers and Inventories	
EC (EINECS/ELINCS)	209-529-3
EC Class	Xi, R: 22-36/37/38, S: 26-36/37/39
RTECS	TS7750000
RTECS class	Mutagen
Merck	12,7781
Beilstein/Gmelin	10012 (G)
EPA OPP	73504
Swiss Giftliste 1	G-1739
Canada DSL/NDSL	DSL
US TSCA	Listed
Austrailia AICS	Listed
New Zealand	Listed
Japan ENCS (MITI)	Listed
Korea ECL	Listed
Philippiens PICCS	Listed

Properties	
Formula	CK2O3
Formula mass	138.21
Melting point, °C	891
Boiling point, °C	(decomposes)
Vapor density (air=1)	AS HOH
Evaporization number	Does not evaporate.
Density	2.42 g/cm^3
Solubility in water	521 g/L (25 C)
Thermal expansion	0.00024/K
Heat of fusion	35.74 kJ/mol

Hazards and Protection	
Storage	Store in a tightly closed container. Store in a cool, dry, well-ventilated area away from incompatible substances.
WHMIS	E.
Handling	Wash thoroughly after handling. Minimize dust generation and accumulation. Avoid contact with eyes, skin, and clothing. Avoid ingestion and inhalation. Use with adequate ventilation.
Protection	Eyes: Wear appropriate protective eyeglasses or chemical safety goggles as described by OSHA's eye and face protection regulations in 29 CFR 1910.133 or European Standard EN166. Skin: Wear appropriate protective gloves to prevent skin exposure. Clothing: Wear appropriate protective clothing to prevent skin exposure.
Respirators	Follow the OSHA respirator regulations found in 29CFR 1910.134 or European Standard EN 149. Always use a NIOSH or European Standard EN 149 approved respirator when necessary.
Small spills/leaks	Clean up spills immediately, using the appropriate protective equipment. Sweep up, then place into a suitable container for disposal. Avoid generating dusty conditions. Provide ventilation.
Stability	Stable under normal shipping and handling conditions.
Incompatibilities	Reacts dangerously with aluminum, fluorine, magnesium, silicon, and chlorine trifluoride.
Decomposition	Carbon monoxide, carbon dioxide, oxides of potassium.

Health	
Poison_Class	4
Exposure effects	No data were available to assess the potential effects of exposure to this agent during pregnancy or lactation.
Ingestion	May cause severe gastrointestinal tract irritation with nausea, vomiting and possible burns. May be harmful if swallowed.
Inhalation	Causes respiratory tract irritation.
Skin	Causes an irritant and caustic action similar to that of potassium hydroxide.
Eyes	Causes eye irritation. Causes an irritant and caustic action similar to that of potassium hydroxide.
First aid	
Ingestion	Do NOT induce vomiting. If victim is conscious and alert, give 2-4 cupfuls of milk or water. Never give anything by mouth to an unconscious person. Get medical aid immediately.
Inhalation	Remove from exposure to fresh air immediately. If not breathing, give artificial respiration. If breathing is difficult, give oxygen. Get medical aid if cough or other symptoms appear.
Skin	Get medical aid. Immediately flush skin with plenty of soap and water for at least 15 minutes while removing contaminated clothing and shoes. Wash clothing before reuse.
Eyes	Immediately flush eyes with plenty of water for at least 15 minutes, occasionally lifting the upper and lower eyelids. Get medical aid immediately.

Transportation	
HS Code	2836 40 00
Std. Transport #	4935230

Potassium Hydroxide	
• Caustic potash • Potassim hydrate • Potassim hydrate • Potassa • Potassium lye	
Formula	KOH
Structure	K^+ OH^-
Description	White flakes.
Uses	Ph adjustment, fungicide, herbicide, microbiocide.

Registry Numbers and Inventories	
CAS	1310-58-3
NIH PubChem CID	14797
EC (EINECS/ELINCS)	215-181-3
EC Index Number	019-002-00-8
EC Class	Xn; R22, C; R35
RTECS	TT2100000
RTECS class	Agricultural Chemical and Pesticide; Mutagen; Primary Irritant
UN (DOT)	1813
Merck	12,7806

(Continued)

Registry Numbers and Inventories	
Beilstein/Gmelin	19033 (G)
EPA OPP	75602
Swiss Giftliste 1	G-4504
Canada DSL/NDSL	DSL
US TSCA	Listed
Austrailia AICS	Listed
New Zealand	Listed
Japan ENCS (MITI)	Listed
Korea ECL	Listed
Philippiens PICCS	Listed
Israel	Listed

Properties	
Formula	HKO
Formula mass	56.11
Melting point, °C	360
Boiling point, °C	1320
Vapor pressure, mm_{Hg}	1.0 (714 C)
Odor threshold	Odorless
Density	2.04 g/cm^3
Solubility in water	1100 g/L
Viscosity	1 cp (550 C)
Refractive index	1.409 - 1.411 (420 C)
pKa/pKb	Strong base
Thermal expansion	0.00019/K
Heat of fusion	6.72 kJ/mol
Heat of vaporization	136.0 kJ/mol

Hazards and Protection	
Storage	Store in a cool, dry place. Keep container closed when not in use. Store in a tightly closed container. Corrosives area.
WHMIS	D1B E
Handling	Wash thoroughly after handling. Do not allow water to get into the container because of violent reaction. Do not breathe dust, vapor, mist, or gas. Do not get in eyes, on skin, or on clothing. Keep container tightly closed. Do not ingest or inhale. Use only in a chemical fume hood. Discard contaminated shoes.
Protection	Eyes: Wear appropriate protective eyeglasses or chemical safety goggles as described by OSHA's eye and face protection regulations in 29 CFR 1910.133 or European Standard EN166. Skin: Wear appropriate protective gloves to prevent skin exposure. Clothing: Wear appropriate protective clothing to prevent skin exposure.
Respirators	A respiratory protection program that meets OSHA's 29 CFR 1910.134 and ANSI Z88.2 requirements or European Standard EN 149 must be followed whenever workplace conditions warrant a respirator's use. Follow the OSHA respirator regulations found in 29CFR 1910.134 or European Standard EN 149. Always use a NIOSH or European Standard EN 149 approved respirator when necessary.
Small spills/leaks	Vacuum or sweep up material and place into a suitable disposal container. Clean up spills immediately, using the appropriate protective equipment. Avoid generating dusty conditions. Provide ventilation.

(*Continued*)

Hazards and Protection	
Stability	Stable under normal shipping and handling conditions. Readily absorbs carbon dioxide and moisture from the air and deliquesces.
Incompatibilities	Halogenated hydrocarbons - halogens - nitrocompounds - organic materials - acid chlorides - acid anhydrides - magnesium - copper - generates large amounts of heat when in contact with water and may steam and splatter. Reacts with chlorine dioxide, nitrobenzene, nitromethane, nitrogen trichloride, peroxidized tetrahydrofuran, 2,4,6-trinitrotoluene, bromoform+ crown ethers, acids alcohols, sugars, germanium cyclopentadiene, maleic dicarbide. Corrosive to metals such as aluminum, tin, and zinc to cause formation of flammable hydrogen gas, moisture.
Decomposition	Oxides of potassium, hydrogen gas.

Fire	
Fire fighting	Wear a self-contained breathing apparatus in pressure-demand, MSHA/NIOSH (approved or equivalent), and full protective gear. During a fire, irritating and highly toxic gases may be generated by thermal decomposition or combustion. Wear appropriate protective clothing to prevent contact with skin and eyes. Wear a self-contained breathing apparatus (SCBA) to prevent contact with thermal decomposition products. Use water with caution and in flooding amounts.

Fire		
		Contact with moisture or water may generate sufficient heat to ignite nearby combustible materials. Containers may explode when heated. Non-combustible, substance itself does not burn but may decompose upon heating to produce irritating, corrosive and/or toxic fumes. Extinguishing media: Use extinguishing media most appropriate for the surrounding fire. DO NOT USE WATER!
Fire potential		Nonflammable.
Hazards		Some are oxidizers and may ignite combustibles (wood, paper, oil, clothing, etc.). Contact with metals may evolve flammable hydrogen gas. Containers may explode when heated.
Combustion products		Non-combustible, substance itself does not burn but may decompose upon heating to produce corrosive and/or toxic fumes.
NFPA	Health	3
	Flammability	0
	Reactivity	1

Health	
Exposure limit(s)	TLV: ppm; 2 mg/m^3 (ceiling values) (ACGIH 1993-1994). NIOSH REL: TWA 2 mg/m^3
Poison_Class	2
Exposure effects	Prolonged or repeated skin contact may cause dermatitis. Prolonged or repeated eye contact may cause conjunctivitis. Effects may be delayed.

(*Continued*)

Health	
Ingestion	Harmful if swallowed. May cause severe and permanent damage to the digestive tract. May cause circulatory system failure. May cause perforation of the digestive tract. Causes severe digestive tract burns with abdominal pain, vomiting, and possible death. May cause systemic effects.
Inhalation	Harmful if inhaled. Irritation may lead to chemical pneumonitis and pulmonary edema. Causes severe irritation of upper respiratory tract with coughing, burns, breathing difficulty, and possible coma. Causes chemical burns to the respiratory tract. Inhalation may be fatal as a result of spasm, inflammation, edema of the larynx and bronchi, chemical pneumonitis and pulmonary edema. Aspiration may lead to pulmonary edema. May cause systemic effects.
Skin	Causes skin burns. May cause deep, penetrating ulcers of the skin. Causes severe burns with delayed tissue destruction. Causes redness and pain. May cause skin rash (in milder cases), and cold and clammy skin with cyanosis or pale color.
Eyes	Causes severe eye burns. May cause irreversible eye injury. Contact may cause ulceration of the conjunctiva and cornea. Eye damage may be delayed. Causes redness and pain. May cause chemical conjunctivitis and corneal damage.

First aid	
Ingestion	Do NOT induce vomiting. If victim is conscious and alert, give 2-4 cupfuls of milk or water. Never give anything by mouth to an unconscious person. Get medical aid immediately.

First aid	
Inhalation	Get medical aid immediately. Remove from exposure to fresh air immediately. If breathing is difficult, give oxygen. DO NOT use mouth-to-mouth respiration. If breathing has ceased apply artificial respiration using oxygen and a suitable mechanical device such as a bag and a mask.
Skin	Get medical aid immediately. Immediately flush skin with plenty of soap and water for at least 15 minutes while removing contaminated clothing and shoes. Wash clothing before reuse. Discard contaminated clothing in a manner which limits further exposure. Destroy contaminated shoes.
Eyes	Immediately flush eyes with plenty of water for at least 15 minutes, occasionally lifting the upper and lower eyelids. Get medical aid immediately. Do NOT allow victim to rub or keep eyes closed. Extensive irrigation is required (at least 30 minutes).

Transportation		
UN number	1813	
Response guide	154	
Hazard class	8	
Packing Group	II	
USCG CHRIS Code	CPS	
USCG Compatatibility Group	5 Caustics	
HS Code	2815 20 10	

(Continued)

Transportation		
Std. Transport #	4932053	
IMO Chemical Code	17	
IMO Pollution Category	C	
IMO Hazard code	S/P	

Potassium Metaborate	
• Boron potassium oxide • Potassium oxido(oxo)borane	
Formula	KBO_2
Structure	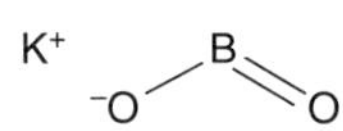
Description	White hexagonal crystals.

Registry Numbers and Inventories	
CAS	13709-94-9
NIH PubChem CID	123325
EC (EINECS/ELINCS)	237-262-2
EC Class	R: 36/37/38, S: 26-36
Beilstein/Gmelin	314006 (G)
Swiss Giftliste 1	G-8892
Canada DSL/NDSL	DSL
US TSCA	Listed
Austrailia AICS	Listed
New Zealand	Listed
Japan ENCS (MITI)	Listed
Korea ECL	Listed

Properties	
Formula	BKO2
Formula mass	82.92
Density	1.76 g/cm^3
Solubility in water	448 g/L (25 C)
Heat of fusion	3.58 kJ/mol

Hazards and Protection	
Storage	Keep tightly closed. Store in a cool dry place.
Protection	Compatible chemical-resistant gloves. Chemical safety goggles.
Respirators	Government approved respirator.
Small spills/leaks	Sweep up, place in a bag and hold for waste disposal. Avoid raising dust. Ventilate area and wash spill site after material pickup is complete.
Incompatibilities	Strong oxidizing agents.
Decomposition	Boron oxides.

Fire	
Fire fighting	Extinguish using water spray, carbon dioxide, dry chemical powder, or appropriate foam. Wear self-contained breathing apparatus and protective clothing to prevent contact with skin and eyes.
Hazards	Emits toxic fumes under fire conditions.

Health	
Poison_Class	3
Exposure effects	Irritant. Irritating to eyes, respiratory system and skin.
Inhalation	Material is irritating to mucous membranes and upper respiratory tract.
First aid	
Ingestion	If swallowed, wash out mouth with water provided person is conscious. Call a physician.

(Continued)

Health	
Inhalation	If inhaled, remove to fresh air. If not breathing give artificial respiration. If breathing is difficult, give oxygen.
Skin	In case of contact, immediately wash skin with soap and copious amounts of water.
Eyes	In case of contact, immediately flush eyes with copious amounts of water for at least 15 minutes.

Quaternary Ammonium Compounds, tri-C(10)-alkylmethyl-, chlorides	
• Adogen 464 • Aliquat 7402 • Aliquat N 263 • Aliquat 336 • Aliquat 336S • Disperstat A	
Formula	Unspecified
Description	Clear almost colorless liquid.

Registry Numbers and Inventories	
CAS	63393-96-4
NIH PubChem CID	3748511 (SID)
EC (EINECS/ELINCS)	264-120-7
RTECS	UZ2997500
RTECS class	Other
UN (DOT)	2924
Beilstein/Gmelin	NA
Canada DSL/NDSL	DSL
US TSCA	Listed
Austrailia AICS	Listed
New Zealand	Listed
Korea ECL	Listed
Philippiens PICCS	Listed

Properties	
Density	0.898 g/cm^3

Hazards and Protection	
Storage	Keep away from heat, sparks, and flame. Keep away from sources of ignition. Keep container closed when not in use. Flammables-area. Store protected from moisture.
Handling	Ground and bond containers when transferring material. Use spark-proof tools and explosion proof equipment. Empty containers retain product residue, (liquid and/or vapor), and can be dangerous. Keep container tightly closed. Avoid contact with heat, sparks and flame. Do not get on skin or in eyes. Do not ingest or inhale. Use with adequate ventilation. Do not allow contact with water. Discard contaminated shoes. Do not pressurize, cut, weld, braze, solder, drill, grind, or expose empty containers to heat, sparks or open flames. Keep from contact with moist air and steam.
Protection	Eyes: Wear appropriate protective eyeglasses or chemical safety goggles as described by OSHA's eye and face protection regulations in 29 CFR 1910.133 or European Standard EN166. Skin: Wear appropriate protective gloves to prevent skin exposure. Clothing: Wear appropriate protective clothing to minimize contact with skin.
Respirators	A respiratory protection program that meets OSHA's 29 CFR 1910.134 and ANSI Z88.2 requirements or European Standard EN 149 must be followed whenever workplace conditions warrant a respirator's use.

(*Continued*)

Hazards and Protection	
Small spills/leaks	Absorb spill with inert material, (e.g., dry sand or earth), then place into a chemical waste container. Clean up spills immediately, using the appropriate protective equipment. Remove all sources of ignition. Use a spark-proof tool. Provide ventilation. Do not get water inside containers. A vapor suppressing foam may be used to reduce vapors.
Stability	Stable under normal temperatures and pressures.
Incompatibilities	Strong oxidizing agents.
Decomposition	Hydrogen chloride, nitrogen oxides, carbon monoxide, irritating and toxic fumes and gases, carbon dioxide, nitrogen.

Fire	
Flash Point,°C	43
Fire fighting	Wear a self-contained breathing apparatus in pressure-demand, MSHA/NIOSH (approved or equivalent), and full protective gear. During a fire, irritating and highly toxic gases may be generated by thermal decomposition or combustion. Will burn if involved in a fire. Use water spray to keep fire-exposed containers cool. Containers may explode in the heat of a fire. Extinguishing media: For small fires, use dry chemical, carbon dioxide, water spray or alcohol-resistant foam. For large fires, use water spray, fog, or alcohol-resistant foam. Use water spray to cool fire-exposed containers. Water may be ineffective. Do NOT get water inside containers. Do NOT use straight streams of water.

Fire		
Fire potential		Flammable/combustible material. May be ignited by heat, sparks or flames.
Hazards		Vapors may form explosive mixtures with air. Vapors may travel to source of ignition and flash back. Most vapors are heavier than air. They will spread along ground and collect in low or confined areas (sewers, basements, tanks). Vapor explosion hazard indoors, outdoors or in sewers.
Combustion products		Fire will produce irritating, corrosive and/or toxic gases.
NFPA	**Health**	3
	Flammability	2
	Reactivity	0

Health	
Exposure effects	
Ingestion	May cause severe and permanent damage to the digestive tract. Causes gastrointestinal tract burns.
Inhalation	Causes chemical burns to the respiratory tract. Vapors may cause dizziness or suffocation.
Skin	Causes skin burns.
Eyes	Causes eye burns.

First aid	
Ingestion	Do NOT induce vomiting. If victim is conscious and alert, give 2-4 cupfuls of milk or water. Never give anything by mouth to an unconscious person. Get medical aid immediately. Do NOT induce vomiting. If conscious and alert, rinse mouth and drink 2-4 cupfuls of milk or water.
Inhalation	Get medical aid immediately. Remove from exposure to fresh air immediately. If not breathing, give artificial respiration. If breathing is difficult, give oxygen. DO NOT use mouth-to-mouth respiration.

(*Continued*)

First aid	
Skin	Get medical aid immediately. Immediately flush skin with plenty of soap and water for at least 15 minutes while removing contaminated clothing and shoes. Wash clothing before reuse. Destroy contaminated shoes.
Eyes	Get medical aid immediately. Do NOT allow victim to rub or keep eyes closed. Extensive irrigation is required (at least 30 minutes).

Transportation		
UN number	2924	
Response guide	132	
Hazard class	3	
Packing Group	I; II; III	

Sodium Carbonate	
• Disodium carbonate • Soda Ash • Trona	
Formula	Na_2CO_3
Structure	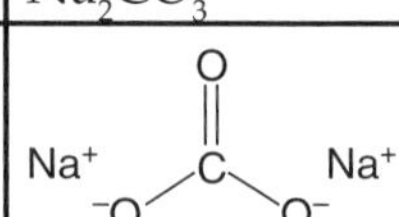
Description	White to greyish crystals, powders or lumps, hygroscopic.
Uses	Baking powder, ph adjustment, fungicide, microbiocide, herbicide.

Registry Numbers and Inventories	
CAS	497-19-8
NIH PubChem CID	10340
EC (EINECS/ELINCS)	207-838-8
EC Index Number	011-005-00-2
EC Class	Xi; R36
RTECS	VZ4050000
RTECS class	Agricultural Chemical and Pesticide; Mutagen; Reproductive Effector; Human Data; Primary Irritant
Merck	12,8739
Beilstein/Gmelin	23964 (G)
EPA OPP	73506
Swiss Giftliste 1	G-2579
Canada DSL/NDSL	DSL
US TSCA	Listed
Austrailia AICS	Listed
New Zealand	Listed
Japan ENCS (MITI)	Listed
Korea ECL	Listed
Philippiens PICCS	Listed

Properties	
Formula	CNa2O3
Formula mass	105.99
Melting point, °C	854
Boiling point, °C	1600
Density	2.45 4 g/cm^3 (18 C)
Solubility in water	365 g/L (20 C)
Viscosity	0.0355 P (900 C)
Surface tension	209 erg/cm^2 (910 C)
Refractive index	1.495
Dielectric constant	8 (18 C).75
Thermal expansion	0.000204/K
Heat of fusion	30.6 kJ/mol

Hazards and Protection	
Storage	Store in a tightly closed container. Store in a cool, dry, well-ventilated area away from incompatible substances.
WHMIS	D1B E
Handling	Wash thoroughly after handling. Remove contaminated clothing and wash before reuse. Do not get in eyes, on skin, or on clothing. Keep container tightly closed. Avoid ingestion and inhalation.
Protection	Eyes: Wear appropriate protective eyeglasses or chemical safety goggles as described by OSHA's eye and face protection regulations in 29 CFR 1910.133 or European Standard EN166. Skin: Wear appropriate protective gloves and clothing to prevent skin exposure. Clothing: Wear appropriate protective clothing to minimize contact with skin.
Respirators	Follow the OSHA respirator regulations found in 29CFR 1910.134 or European Standard EN 149. Always use a NIOSH or European Standard EN 149 approved respirator when necessary.
Small spills/leaks	Vacuum or sweep up material and place into a suitable disposal container. Avoid generating dusty conditions. Provide ventilation. Cover with material such as dry soda ash or calcium carbonate and place into a closed container for disposal.
Disposal code	14
Stability	Stable at room temperature in closed containers under normal storage and handling conditions.

Hazards and Protection	
Incompatibilities	Reacts explosively with red-hot alumium metal. Incompatible with ammonia + silver nitrate, 2,4-dinitrotoluene, 2,4,6-trinitrotoluene, sulfuric acid, sodium sulfide + water, lithium, phosphorus pentoxide, fluorine, and hydrogen peroxide. Hot concentrated solutions of sodium carbonate are mildly corrosive to steel.
Decomposition	Carbon dioxide, toxic fumes of sodium oxide.

Fire		
Fire fighting		Wear a self-contained breathing apparatus in pressure-demand, MSHA/NIOSH (approved or equivalent), and full protective gear. Material will not burn. Wear appropriate protective clothing to prevent contact with skin and eyes. Wear a self-contained breathing apparatus (SCBA) to prevent contact with thermal decomposition products. Extinguishing media: Substance is noncombustible; use agent most appropriate to extinguish surrounding fire. In case of fire use water fog, dry chemical, carbon dioxide or alcohol type foam.
NFPA	**Health**	3
	Flammability	0
	Reactivity	0

Health	
Poison_Class	5
Exposure effects	Prolonged or repeated inhalation may cause nosebleeds, nasal congestion, erosion of the teeth, perforation of the nasal septum, chest pain and bronchitis.
Ingestion	May cause irritation of the digestive tract.
Inhalation	Harmful if inhaled. May cause irritation of the respiratory tract with burning pain in the nose and throat, coughing, wheezing, shortness of breath and pulmonary edema.
Skin	Contact with skin causes irritation and possible burns, especially if the skin is wet or moist.
Eyes	May result in corneal injury. Contact with eyes may cause severe irritation, and possible eye burns.
First aid	
Ingestion	Do NOT induce vomiting. If victim is conscious and alert, give 2-4 cupfuls of milk or water. Never give anything by mouth to an unconscious person. Get medical aid immediately.
Inhalation	Remove from exposure to fresh air immediately. If not breathing, give artificial respiration. If breathing is difficult, give oxygen. Get medical aid if cough or other symptoms appear.
Skin	Get medical aid. Flush skin with plenty of soap and water for at least 15 minutes while removing contaminated clothing and shoes. Wash clothing before reuse.
Eyes	Immediately flush eyes with plenty of water for at least 15 minutes, occasionally lifting the upper and lower eyelids. Get medical aid immediately.

Transportation	
USCG CHRIS Code	SCE
USCG Compatatibility Group	5 Caustics
HS Code	2836 20 00
IMO Chemical Code	18
IMO Pollution Category	D

Sodium Chloride	
• Table salt • Rock salt • Halite	
Formula	NaCl
Structure	Na^+ Cl^-
Description	Colorless, transparent crystals or white crystalline powder, odorless, hygroscopic.
Uses	In the production of chemicals (sodium hydroxide, soda ash, hydrogen chloride, chlorine, metallic sodium), ceramic glazes, metallurgy, curing hides, food preservative, mineral waters, soap manufacture (salting out), home water softeners, highway deicing, regeneration of ion-exchange resins, photography, food seasoning, herbicide, fire extinguishing, nuclear reactors, mouthwash, medicine (heat exhaustion), salting out dyestuffs, supercooled solutions. Single crystals are used for spectroscopy, uv and infrared transmissions.

Registry Numbers and Inventories	
CAS	7647-14-5
NIH PubChem CID	5234
EC (EINECS/ELINCS)	231-598-3
EC Class	R: 36, S: 24/25
RTECS	VZ4725000
RTECS class	Agricultural Chemical and Pesticide; Mutagen; Reproductive Effector; Human Data; Primary Irritant
Merck	12,8742
Beilstein/Gmelin	13673 (G)
EPA OPP	13905
Swiss Giftliste 1	G-2580
Canada DSL/NDSL	DSL
US TSCA	Listed
Austrailia AICS	Listed
New Zealand	Listed
Japan ENCS (MITI)	Listed
Korea ECL	Listed
Philippiens PICCS	Listed

Properties	
Formula	NaCl
Formula mass	58.44
Melting point, °C	801
Boiling point, °C	1413
Vapor pressure, mm_{Hg}	9.01575 (1026.85 C)
Saturation Concentration	Approximately zero
Density	2.16 g/cm^3
Solubility in water	360 g/L
Thermal expansion	0.000118 - 0.000232/K
Heat of fusion	28.16 kJ/mol

Hazards and Protection	
Storage	Store in a cool, dry, well-ventilated area away from incompatible substances.
Handling	Use with adequate ventilation. Minimize dust generation and accumulation. Avoid contact with eyes, skin, and clothing. Keep container tightly closed. Do not ingest or inhale.
Protection	Eyes: Wear safety glasses with side shields. Skin: Wear appropriate gloves to prevent skin exposure. Clothing: Wear appropriate protective clothing to minimize contact with skin.
Respirators	A NIOSH/MSHA approved air purifying dust or mist respirator or European Standard EN 149.
Small spills/leaks	Vacuum or sweep up material and place into a suitable disposal container. Clean up spills immediately, using the appropriate protective equipment. Avoid generating dusty conditions.
Disposal code	14
Stability	Stable under normal shipping and handling conditions.
Incompatibilities	Reacts with most nonnoble metals such as iron or steel, building materials (such as cement), bromine, or trifluoride. Potentially explosive reaction with dichloromaleic anhydride + urea. Electrolysis of mixtures with nitrogen compounds may form explosive nitrogen trichloride.
Decomposition	Toxic fumes of sodium oxide.

Fire		
Fire fighting		Fires involving this chemical can be controlled with a dry chemical, carbon dioxide, foam or Halon extinguisher.
NFPA	Health	1
	Flammability	0
	Reactivity	0

Health	
Poison_Class	-
Exposure effects	Hypernatremia may produce restlessness, irritability, dizziness, progressive obtundation, reduced salivation, tearing, seizures and coma. Hyperosmolarity of cerebral fluids may lead to cellular crenation and irreversible neurologic damage.
Ingestion	Ingestion of large amounts may cause gastrointestinal irritation. Ingestion of large amounts may cause nausea and vomiting, rigidity or convulsions. Continued exposure can produce coma, dehydration, and internal organ
Inhalation	May cause respiratory tract irritation.
Skin	May cause skin irritation.
Eyes	May cause eye irritation.

First aid	
Ingestion	If victim is conscious and alert, give 2-4 cupfuls of milk or water. Get medical aid. Wash mouth out with water.
Inhalation	Remove from exposure to fresh air immediately. If not breathing, give artificial respiration. If breathing is difficult, give oxygen. Get medical aid if cough or other symptoms appear.

First aid	
Skin	Flush skin with plenty of soap and water for at least 15 minutes while removing contaminated clothing and shoes. Get medical aid if irritation develops or persists. Wash clothing before reuse.
Eyes	Flush eyes with plenty of water for at least 15 minutes, occasionally lifting the upper and lower eyelids. Get medical aid.

Transportation	
HS Code	3822 00 00

Sodium Erythorbate	
• Sodium isoascorbate • D-(+)-Isoascorbic acid, sodium salt • Erythorbic acid sodium salt • D-Erythro-hex-2-enonic acid, g-lactone, monosodium salt • Neo-Cebitate • Sodium Isovitamin C	
Formula	$C_6H_7O_6$. Na
Structure	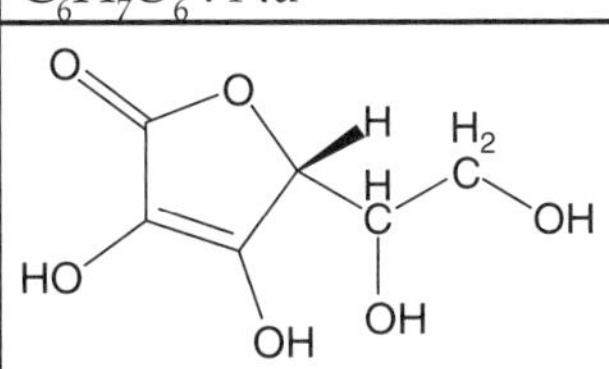
Description	Almost odorless fluffy, white to off-white crystalline powder.
Uses	Preservative.

Registry Numbers and Inventories	
CAS	**6381-77-7**
NIH PubChem CID	**23687319**
EC (EINECS/ELINCS)	**228-973-9**

(Continued)

Registry Numbers and Inventories	
RTECS	MP8910000
RTECS class	Primary Irritant
Beilstein/Gmelin	5465157
EPA OPP	116401
Canada DSL/NDSL	DSL
US TSCA	Listed
Austrailia AICS	Listed
New Zealand	Listed
Japan ENCS (MITI)	Listed
Korea ECL	Listed
Philippiens PICCS	Listed

Properties	
Formula	C6H7NaO6
Formula mass	198.11
Melting point, °C	168-170
Density	1.2 g/cm^3 (20 C)
Solubility in water	153 g/L

Hazards and Protection	
Storage	Store in a cool, dry place. Store in a tightly closed container.
Handling	Wash thoroughly after handling. Remove contaminated clothing and wash before reuse. Use with adequate ventilation. Avoid contact with skin and eyes. Avoid ingestion and inhalation.
Protection	Eyes: Wear appropriate protective eyeglasses or chemical safety goggles as described by OSHA's eye and face protection regulations in 29 CFR 1910.133 or European Standard EN166. Skin: Wear appropriate protective gloves to prevent skin exposure. Clothing: Wear appropriate protective clothing to prevent skin exposure.

Hazards and Protection	
Respirators	Follow the OSHA respirator regulations found in 29CFR 1910.134 or European Standard EN 149. Always use a NIOSH or European Standard EN 149 approved respirator when necessary.
Small spills/leaks	Vacuum or sweep up material and place into a suitable disposal container.
Stability	Stable under normal temperatures and pressures.
Incompatibilities	Strong oxidizing agents, strong bases, chemically active metals, aluminum, zinc, magnesium,copper.
Decomposition	Carbon monoxide, carbon dioxide.

Fire	
Fire fighting	Wear a self-contained breathing apparatus in pressure-demand, MSHA/NIOSH (approved or equivalent), and full protective gear. To extinguish fire use water spray, dry chemical, carbon dioxide, or chemical foam.
Fire potential	This material is probably combustible.

Health	
Exposure effects	
Ingestion	May cause irritation of the digestive tract. The toxicological properties of this substance have not been fully investigated.
Inhalation	May cause respiratory tract irritation. The toxicological properties of this substance have not been fully investigated.
Skin	May cause skin irritation.
Eyes	May cause eye irritation.
First aid	
Ingestion	If victim is conscious and alert, give 2-4 cupfuls of milk or water. Get medical aid. Wash mouth out with water.

(*Continued*)

Health	
Inhalation	Get medical aid immediately. Remove from exposure to fresh air immediately. If not breathing, give artificial respiration. If breathing is difficult, give oxygen.
Skin	Get medical aid. Flush skin with plenty of soap and water for at least 15 minutes while removing contaminated clothing and shoes.
Eyes	Flush eyes with plenty of water for at least 15 minutes, occasionally lifting the upper and lower eyelids. Get medical aid.

Sodium Hydroxide	
• Caustic soda • Soda lye • Lye • Sodium hydrate	
Formula	NaOH
Structure	Na^+ OH^-
Description	White, deliquescent pellets or flakes. Hygroscopic.
Uses	Vegetable oil refining, regenerating ion exchange resins, organic fusions, peeling of fruits and vegetables in food industry, etching and electroplating.

Registry Numbers and Inventories	
CAS	1310-73-2
NIH PubChem CID	14798
EC (EINECS/ELINCS)	215-185-5
EC Index Number	011-002-00-6
EC Class	C; R35
RTECS	WB4900000
RTECS class	Agricultural Chemical and Pesticide; Mutagen; Primary Irritant

Registry Numbers and Inventories	
UN (DOT)	1823
Merck	12,8772
Beilstein/Gmelin	19031 (G)
EPA OPP	75603
Swiss Giftliste 1	G-4492
Canada DSL/NDSL	DSL
US TSCA	Listed
Austrailia AICS	Listed
New Zealand	Listed
Japan ENCS (MITI)	Listed
Korea ECL	Listed
Philippiens PICCS	Listed
Israel	Listed

Properties	
Formula	HNaO
Formula mass	40.00
Melting point, °C	323
Boiling point, °C	1390
Vapor pressure, mm_{Hg}	0.07 (300 C)
Saturation Concentration	1980 ppm (0.2%) at 20 C (50 % solution) (calculated)
Odor threshold	Odorless
Density	2.13 g/cm^3
Solubility in water	1.1 kg/L
Viscosity	4.0 cp at 350C.
Surface tension	74.35 g/s2 at 18 C
Refractive index	1.412 (420 C)
Dielectric constant	57.5 (25 C)
pKa/pKb	Strong base
Thermal expansion	8.4E-5/K (20 C)
Heat of fusion	8.4 kJ/mol
Heat of vaporization	175 kJ/mol

Hazards and Protection	
Storage	Store in a tightly closed container. Store in a cool, dry, well-ventilated area away from incompatible substances. Corrosives area. Store protected from moisture. Do not store with aluminum or magnesium. Do not mix with acids or organic materials.
WHMIS	E
Handling	Wash thoroughly after handling. Do not allow water to get into the container because of violent reaction. Minimize dust generation and accumulation. Avoid contact with eyes, skin, and clothing. Keep container tightly closed. Avoid ingestion and inhalation. Use with adequate ventilation. Discard contaminated shoes.
Protection	Eyes: Wear appropriate protective eyeglasses or chemical safety goggles as described by OSHA's eye and face protection regulations in 29 CFR 1910.133 or European Standard EN166. Skin: Wear appropriate protective gloves to prevent skin exposure. Clothing: Wear appropriate protective clothing to minimize contact with skin.
Respirators	If the exposure limit is exceeded and engineering controls are not feasible, a half facepiece particulate respirator (NIOSH type N95 or better filters) may be worn for up to ten times the exposure limit or the maximum use concentration specified by the appropriate regulatory agency or respirator supplier, whichever is lowest.. A full-face piece particulate respirator (NIOSH type N100 filters) may be worn up to 50 times the exposure limit, or the maximum use concentration specified by the appropriate regulatory agency, or respirator supplier, whichever is lowest.

Hazards and Protection	
	If oil particles (e.g. lubricants, cutting fluids, glycerine, etc.) are present, use a NIOSH type R or P filter. For emergencies or instances where the exposure levels are not known, use a full-facepiece positive-pressure, air-supplied respirator.
Small spills/leaks	Vacuum or sweep up material and place into a suitable disposal container. Avoid runoff into storm sewers and ditches which lead to waterways. Clean up spills immediately, using the appropriate protective equipment. Avoid generating dusty conditions. Provide ventilation.
Stability	Stable at room temperature in closed containers under normal storage and handling conditions.
Incompatibilities	Sodium hydroxide in contact with acids and organic halogen compounds, especially trichloroethylene, may causes violent reactions. Contact with nitromethane and other similar nitro compounds causes formation of shock-sensitive salts. Contact with metals such as aluminum, magnesium, tin, and zinc cause formation of flammable hydrogen gas. Sodium hydroxide, even in fairly dilute solution, reacts readily with various sugars to produce carbon monoxide
Decomposition	Irritating and toxic fumes and gases, toxic fumes of sodium oxide, sodium peroxide fumes.

Fire		
Fire fighting		In the event of a fire, wear full protective clothing and NIOSH-approved self-contained breathing apparatus with full facepiece operated in the pressure demand or other positive pressure mode. Use any means suitable for extinguishing surrounding fire. Adding water to caustic solution generates large amounts of heat.
Fire potential		Nonflammable. Not considered to be a fire hazard. Hot or molten material can react violently with water. Can react with certain metals, such as aluminum, to generate flammable hydrogen gas.
Hazards		Some are oxidizers and may ignite combustibles (wood, paper, oil, clothing, etc.). Contact with metals may evolve flammable hydrogen gas. Containers may explode when heated.
Combustion products		Non-combustible, substance itself does not burn but may decompose upon heating to produce corrosive and/or toxic fumes.
NFPA	**Health**	4
	Flammability	0
	Reactivity	4

Health	
Exposure limit(s)	OSHA PEL: TWA 2 mg/m^3 NIOSH REL: C 2 mg/m^3 NIOSH IDLH: 10 mg/m^3
Poison_Class	4
Exposure effects	Prolonged or repeated skin contact may cause dermatitis. Effects may be delayed.

Health	
Ingestion	May cause severe and permanent damage to the digestive tract. Causes gastrointestinal tract burns. May cause perforation of the digestive tract. Causes severe pain, nausea, vomiting, diarrhea, and shock. May cause corrosion and permanent tissue destruction of the esophagus and digestive tract. May cause systemic effects.
Inhalation	Irritation may lead to chemical pneumonitis and pulmonary edema. Causes severe irritation of upper respiratory tract with coughing, burns, breathing difficulty, and possible coma. Causes chemical burns to the respiratory tract. Aspiration may lead to pulmonary edema. May cause systemic effects.
Skin	Causes skin burns. May cause deep, penetrating ulcers of the skin. May cause skin rash (in milder cases), and cold and clammy skin with cyanosis or pale color.
Eyes	Causes eye burns. May cause chemical conjunctivitis and corneal damage.

First aid	
Ingestion	Do NOT induce vomiting. If victim is conscious and alert, give 2-4 cupfuls of milk or water. Get medical aid immediately.
Inhalation	Get medical aid immediately. Remove from exposure to fresh air immediately. If not breathing, give artificial respiration. If breathing is difficult, give oxygen. DO NOT use mouth-to-mouth respiration. If breathing has ceased apply artificial respiration using oxygen and a suitable mechanical device such as a bag and a mask.

(*Continued*)

First aid	
Skin	Get medical aid. Flush skin with plenty of soap and water for at least 15 minutes while removing contaminated clothing and shoes. Wash clothing before reuse. Discard contaminated clothing in a manner which limits further exposure.
Eyes	Get medical aid. Do NOT allow victim to rub or keep eyes closed. Extensive irrigation is required (at least 30 minutes).

Transportation		
UN number	1823	
Response guide	154	
Hazard class	8	
Packing Group	II	
USCG CHRIS Code	CSS	
USCG Compatatibility Group	5 Caustics	
HS Code	2815 11 00	
Std. Transport #	4931314	
IMO Chemical Code	17	
IMO Pollution Category	D	
IMO Hazard code	S	

Sodium Lauryl Sulfate
• Dodecyl sodium sulfate • Lauryl sodium sulfate • Sodium dodecyl sulfate • Irium

Sodium Lauryl Sulfate	
Formula	$C_{12}H_{25}OSO_2ONa$
Structure	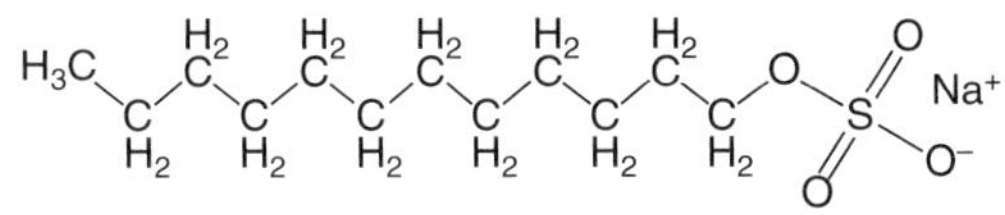
Description	White, cream-colored or light yellow crystals, flakes or powder. Faint fatty odor.
Uses	Surface-active agent for emulsion polymerization, in metal processing, detergents & shampoo, emulsifying, foaming, wetting, dispersing agent in creams, lotion & medical preparations, foaming, wetting & dispersing agent in toothpaste, emulsifier, whipping agent & surfactant in foods.

Registry Numbers and Inventories	
CAS	151-21-3
NIH PubChem CID	3423265
EC (EINECS/ELINCS)	205-788-1
EC Class	Xn; Xi, R: 22-36/37/38, S: 26-36/37/39
RTECS	WT1050000
RTECS class	Drug; Mutagen; Reproductive Effector; Primary Irritant
Merck	12,8782
Beilstein/Gmelin	3599286
EPA OPP	79011
FEMA	4437
Swiss Giftliste 1	G-2596
Canada DSL/NDSL	DSL
US TSCA	Listed
Austrailia AICS	Listed
New Zealand	Listed
Japan ENCS (MITI)	Listed
Korea ECL	Listed
Philippiens PICCS	Listed

Properties	
Formula	$C_{12}H_{25}NaO_4S$
Formula mass	288.38
Melting point, °C	204 - 207
Decomposition point, °C	380
Density	1.1 g/cm^3 (20 C)
Solubility in water	100 g/L
Surface tension	28 g/s2 (25 C)
Partition coefficient, pK_{ow}	1.60

Hazards and Protection	
Storage	Store in a cool, dry place. Store in a tightly closed container.
Handling	Wash thoroughly after handling. Wash hands before eating. Use only in a well ventilated area. Do not breathe dust, vapor, mist, or gas. Do not get on skin or in eyes. Do not ingest or inhale.
Protection	Eyes: Wear appropriate protective eyeglasses or chemical safety goggles as described by OSHA's eye and face protection regulations in 29 CFR 1910.133 or European Standard EN166. Skin: Wear appropriate protective gloves to prevent skin exposure. Clothing: Wear appropriate protective clothing to prevent skin exposure.
Respirators	Follow the OSHA respirator regulations found in 29CFR 1910.134 or European Standard EN 149. Always use a NIOSH or European Standard EN 149 approved respirator when necessary.
Small spills/leaks	Vacuum or sweep up material and place into a suitable disposal container. Reduce airborne dust and prevent scattering by moistening with water. Clean up spills immediately, using the appropriate protective equipment.

Hazards and Protection	
Stability	Stable under normal temperatures and pressures.
Incompatibilities	Strong oxidizing agents.
Decomposition	Carbon monoxide, oxides of sulfur, carbon dioxide.

Fire		
Flash Point,°C		>150
Fire fighting		Fires involving this material can be controlled with a dry chemical, carbon dioxide or Halon extinguisher. A water spray may also be used.
Fire potential		This material is probably combustible.
NFPA	**Health**	2
	Flammability	0
	Reactivity	0

Health	
Poison_Class	4
Exposure effects	Repeated exposure may cause allergic respiratory reaction (asthma). Chronic inhalation may lead to decreased pulmonary function.
Ingestion	Causes gastrointestinal irritation with nausea, vomiting and diarrhea.
Inhalation	Causes respiratory tract irritation. May cause severe allergic respiratory reaction. May cause severe irritation of the respiratory tract with sore throat, coughing, shortness of breath and delayed lung edema.
Skin	Causes skin irritation.
Eyes	Causes eye irritation.

First aid	
Ingestion	Do NOT induce vomiting. If victim is conscious and alert, give 2-4 cupfuls of milk or water. Never give anything by mouth to an unconscious person. Get medical aid.
Inhalation	Get medical aid immediately. Remove from exposure to fresh air immediately. If not breathing, give artificial respiration. If breathing is difficult, give oxygen.
Skin	Immediately flush skin with plenty of soap and water for at least 15 minutes while removing contaminated clothing and shoes. Get medical aid if irritation develops or persists.
Eyes	Immediately flush eyes with plenty of water for at least 15 minutes, occasionally lifting the upper and lower eyelids. Get medical aid.

Transportation	
USCG CHRIS Code	DDS
HS Code	2920 90 10

Sodium Tetraborate	
• Sodium pyroborate • Fused borax • Disodium tetraborate, anhydrous	
Formula	$Na_2B_4O_7$
Structure	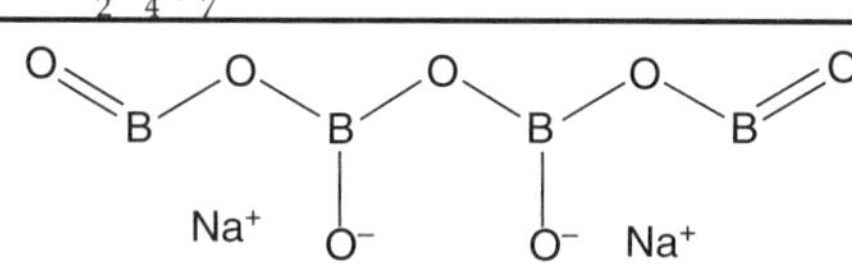
Description	White, odorless and crystalline, hygroscopic.
Uses	Manufacture of glass, enamels, & other ceramic products.

Registry Numbers and Inventories	
CAS	1330-43-4
NIH PubChem CID	10219853
EC (EINECS/ELINCS)	215-540-4
EC Index Number	005-007-00-2
EC Class	S: 24/25
RTECS	ED4588000
RTECS class	Reproductive Effector
Merck	12,8733
Beilstein/Gmelin	35010 (G)
EPA OPP	11112
Swiss Giftliste 1	G-82384
Canada DSL/NDSL	DSL
US TSCA	Listed
Austrailia AICS	Listed
New Zealand	Listed
Japan ENCS (MITI)	Listed
Korea ECL	Listed
Philippiens PICCS	Listed

Properties	
Formula	B4Na2O7
Formula mass	201.22
Melting point, °C	742
Boiling point, °C	1575
Vapor pressure, mm_{Hg}	13 (1250 C)
Saturation Concentration	Approximately zero
Density	2.38 g/cm^3
Solubility in water	26 g/L
Thermal expansion	0.00021/K

Hazards and Protection	
Storage	Store in a cool, dry place. Keep container closed when not in use.
Handling	Wash thoroughly after handling. Remove contaminated clothing and wash before reuse. Minimize dust generation and accumulation. Avoid contact with eyes, skin, and clothing. Avoid ingestion and inhalation.
Protection	Eyes: Wear appropriate protective eyeglasses or chemical safety goggles as described by OSHA's eye and face protection regulations in 29 CFR 1910.133 or European Standard EN166. Skin: Wear appropriate protective gloves to prevent skin exposure. Clothing: Wear appropriate protective clothing to prevent skin exposure.
Respirators	Follow the OSHA respirator regulations found in 29CFR 1910.134 or European Standard EN 149. Always use a NIOSH or European Standard EN 149 approved respirator when necessary.
Small spills/leaks	Clean up spills immediately, using the appropriate protective equipment. Sweep up, then place into a suitable container for disposal. Avoid generating dusty conditions. Provide ventilation.
Disposal code	14
Stability	Stable under normal shipping and handling conditions. Stable under normal temperatures and pressures.
Incompatibilities	Incompatible with alkaloidol salts, mercuric chloride, zinc sulfate, and other metallic salts. Avoid contact with moisture.
Decomposition	Irritating and toxic fumes and gases, toxic fumes of sodium oxide, oxides of boron.

Fire		
Fire fighting		Wear a self-contained breathing apparatus in pressure-demand, MSHA/NIOSH (approved or equivalent), and full protective gear. During a fire, irritating and highly toxic gases may be generated by thermal decomposition or combustion. Non-combustible, substance itself does not burn but may decompose upon heating to produce irritating, corrosive and/or toxic fumes. To extinguish fire, use water, dry chemical, chemical foam, or alcohol-resistant foam. Substance is noncombustible; use agent most appropriate to extinguish surrounding fire.
Fire potential		Not flammable
NFPA	Health	2
	Flammability	0
	Reactivity	0

Health	
Exposure limit(s)	TLV: ppm; 1 mg/m^3 (as TWA) (ACGIH 1994-1995).
Poison_Class	4
Exposure effects	Significant ingestions or dermal exposures can be associated with weak, rapid pulse, cyanosis and abnormally low blood pressure. The patient may present with reduced body temperature, elevated body temperature or normal body temperature. Headache, lethargy, restlessness, weakness, CNS irritation, and/or seizures may occur with long term or repeated exposures. There is insufficient information concerning the reproductive effects of borates in humans. Adverse testicular effects and infertility have been reported in animals.

(Continued)

Health	
	 There have been limited animal studies which suggest decreased ovulation, fetotoxicity and developmental defects may occur with very high exposure levels. Maternal toxicity was present in some studies.
Ingestion	Symptoms may include: headache, excitement, fatigue, nausea, vomiting, stupor, and coma. May cause kidney damage. May cause severe gastrointestinal tract irritation with nausea, vomiting and possible burns. May cause central nervous system depression, characterized by excitement, followed by headache, dizziness, drowsiness, and nausea. Advanced stages may cause collapse, unconsciousness, coma and possible death due to respiratory failure.
Inhalation	Low hazard for usual industrial handling.
Skin	May cause skin irritation. Low hazard for usual industrial handling.
Eyes	Dust may cause mechanical irritation.
First aid	
Ingestion	Never give anything by mouth to an unconscious person. Get medical aid immediately. Do NOT induce vomiting. If conscious and alert, rinse mouth and drink 2-4 cupfuls of milk or water.
Inhalation	Remove from exposure to fresh air immediately. If not breathing, give artificial respiration. If breathing is difficult, give oxygen. Get medical aid if cough or other symptoms appear.
Skin	Get medical aid. Flush skin with plenty of soap and water for at least 15 minutes while removing contaminated clothing and shoes. Wash clothing before reuse.
Eyes	Flush eyes with plenty of water for at least 15 minutes, occasionally lifting the upper and lower eyelids. Get medical aid.

Transportation	
HS Code	2840 11 00

Thioglycolic Acid	
• Mercaptoacetic acid • Acetyl mercaptan • Mercaptoacetate • 2-Thioglycolic acid • 2-Mercaptoacetic acid • 2-Sulfanylacetic acid	
Formula	$HSCH_2COOH$
Structure	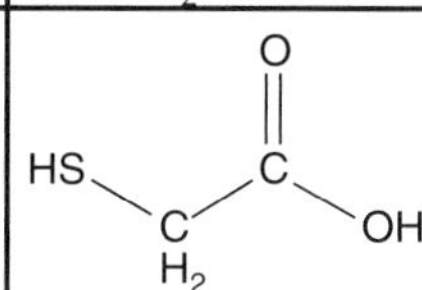
Description	Colorless liquid with a characteristic strong unpleasant odor, lachrymator.
Uses	Chemical intermediate for thioglycolates (eg, calcium thioglycolate), ingredient of depilatories, permanent hair wave solutions and biological media for microorganism growth, reagent for detection of iron and other metal ions, chelating agent.

Registry Numbers and Inventories.	
CAS	68-11-1
NIH PubChem CID	1133
EC (EINECS/ELINCS)	200-677-4
EC Index Number	607-090-00-6
EC Class	T; R23/24/25, C; R34
RTECS	AI5950000

RTECS class	Agricultural Chemical and Pesticide; Primary Irritant
UN (DOT)	1940
Merck	12,9472
Beilstein/Gmelin	506166
Beilstein Reference	4-03-00-00600
Swiss Giftliste 1	G-2940
Canada DSL/NDSL	DSL
US TSCA	Listed
Austrailia AICS	Listed
New Zealand	Listed
Japan ENCS (MITI)	Listed
Korea ECL	Listed
Philippiens PICCS	Listed

Properties	
Formula	$C_2H_4O_2S$
Formula mass	92.11
Melting point, °C	-16.5
Boiling point, °C	96
Vapor pressure, mm_{Hg}	0.03 (25 C)
Vapor density (air=1)	3.18
Saturation Concentration	13200 ppm (1.3%) at 18 C (calc.)
Density	1.326 g/cm^3 (17 C)
Solubility in water	Miscible
Viscosity	6.55 cp (=cp) @ 20C
Refractive index	1.503 (20 C)
pKa/pKb	3.73 (pKa)
Partition coefficient, pK_{ow}	0.09
Heat of vaporization	50.9 kJ/mol
Heat of combustion	-1450 kJ/mol

Hazards and Protection	
Storage	Keep away from heat and flame. Do not store near combustible materials. Store in a cool, dry place. Keep container closed when not in use. Keep away from metals. Corrosives area. Do not store in metal containers.
WHMIS	D1B E
Handling	Wash thoroughly after handling. Use with adequate ventilation. Do not allow water to get into the container because of violent reaction. Do not reuse this container. Empty containers retain product residue, (liquid and/or vapor), and can be dangerous. Do not get on skin or in eyes. Do not ingest or inhale. Keep away from clothing and other combustible materials. Discard contaminated shoes. Do not pressurize, cut, weld, braze, solder, drill, grind, or expose empty containers to heat, sparks or open flames.
Protection	Eyes: Wear appropriate protective eyeglasses or chemical safety goggles as described by OSHA's eye and face protection regulations in 29 CFR 1910.133 or European Standard EN166. Skin: Wear appropriate protective gloves to prevent skin exposure. Clothing: Wear appropriate protective clothing to prevent skin exposure.
Respirators	A respiratory protection program that meets OSHA's 29 CFR 1910.134 and ANSI Z88.2 requirements or European Standard EN 149 must be followed whenever workplace conditions warrant a respirator's use.

(*Continued*)

Hazards and Protection	
Small spills/leaks	Absorb spill with inert material, (e.g., dry sand or earth), then place into a chemical waste container. Avoid runoff into storm sewers and ditches which lead to waterways. Clean up spills immediately, using the appropriate protective equipment. Remove all sources of ignition. Provide ventilation. Do not get water inside containers.
Stability	Stable at room temperature in closed containers under normal storage and handling conditions. Oxidizes when exposed to air.
Incompatibilities	Strong oxidizing agents and strong bases. Reacts with group 1 and 2 metals (Sodium, Potassium, Magnesium, ...) to produce hydrogen gas. Corrosive to steel, stainless steel, nickel base alloys, and aluminum.
Decomposition	Carbon monoxide, irritating and toxic fumes and gases, carbon dioxide, sulfur oxides (SOx), including sulfur oxide and sulfur dioxide.

Fire	
Flash Point,°C	132
Autoignition, °C	350
Lower exp. limit, %	5.9

Fire		
Fire fighting		Wear a self-contained breathing apparatus in pressure-demand, MSHA/NIOSH (approved or equivalent), and full protective gear. During a fire, irritating and highly toxic gases may be generated by thermal decomposition or combustion. Reacts with most metals to form highly flammable hydrogen gas which can form explosive mixtures with air. Reacts violently with water. May polymerize explosively when involved in a fire. Containers may explode when heated. Extinguishing media: Do NOT get water inside containers. For small fires, use dry chemical, carbon dioxide, or water spray. For large fires, use dry chemical, carbon dioxide, alcohol-resistant foam, or water spray. Cool containers with flooding quantities of water until well after fire is out.
Fire potential		May burn but does not ignite readily.
Hazards		When heated, vapors may form explosive mixtures with air: indoors, outdoors, and sewers explosion hazards. May polymerize explosively when heated or involved in a fire. Contact with metals may evolve flammable hydrogen gas. Containers may explode when heated. Runoff may pollute waterways. Substance may be transported in a molten form.
Combustion products		Fire may produce irritating, corrosive and/or toxic gases.
NFPA	Health	3
	Flammability	0
	Reactivity	2

Health	
Exposure limit(s)	NIOSH REL: TWA 1 ppm (4 mg/m^3) [skin]
Poison_Class	3
Exposure effects	CNS depression, seizures, and paralysis were seen in animal experiments, but not in human exposures.
Ingestion	Harmful if swallowed. Causes gastrointestinal tract burns. Mercaptans may cause nausea and headache. Exposure to high concentrations of mercaptans can produce unconsciousness with cyanosis, cold extremities and rapid pulse.
Inhalation	May cause severe irritation of the respiratory tract with sore throat, coughing, shortness of breath and delayed lung edema. Olfactory fatigue may occur. Exposure to high concentrations of mercaptans can produce unconsciousness with cyanosis, cold extremities and rapid pulse. Mercaptans may cause nausea and headache.
Skin	May be fatal if absorbed through the skin. Causes skin burns. Strong irritant of the skin.
Eyes	Causes severe eye burns. May cause irreversible eye injury.

First aid	
Ingestion	Do NOT induce vomiting. If victim is conscious and alert, give 2-4 cupfuls of milk or water. Never give anything by mouth to an unconscious person. Get medical aid immediately.
Inhalation	Get medical aid immediately. Remove from exposure to fresh air immediately. If not breathing, give artificial respiration. If breathing is difficult, give oxygen. DO NOT use mouth-to-mouth respiration.
Skin	Get medical aid immediately. Flush skin with plenty of soap and water for at least 15 minutes while removing contaminated clothing and shoes. Wash clothing before reuse. Destroy contaminated shoes.

First aid	
Eyes	Get medical aid immediately. Do NOT allow victim to rub or keep eyes closed. Extensive irrigation is required (at least 30 minutes).

Transportation		
UN number	1940	
Response guide	153	
Hazard class	8	
Packing Group	II	
HS Code	2930 90 70	
Std. Transport #	4931460	

Triethanolamine Lauryl Sulfate	
• Lauryl sulfate triethanolamine salt • Triethanolamine dodecyl sulfate • Sulfuric acid, monododecyl ester, compd. with 2,2′,2″-nitrilotris[ethanol] (1:1) • TEA-lauryl sulfate • Drene • 2-(Bis(2-hydroxyethyl)amino)ethanol; dodecyl hydrogen sulfate	
Formula	$C_{12}H_{26}O_4S.C_6H_{15}NO_3$
Structure	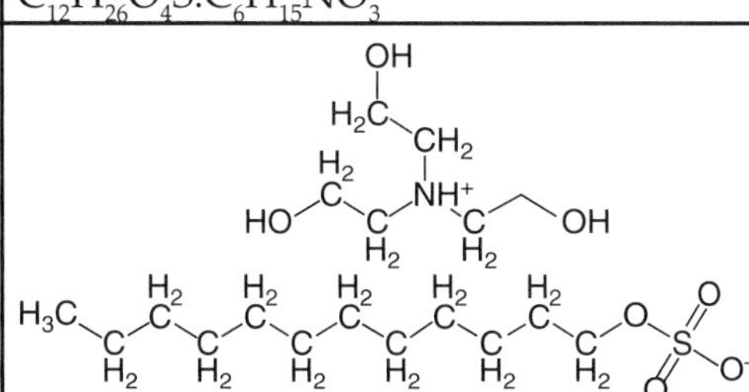
Description	Clear, pale yellow liquid.
Uses	Surfactant for fire fighting foams, textile auxiliary.

Registry Numbers and Inventories	
CAS	139-96-8
NIH PubChem CID	8777
EC (EINECS/ELINCS)	205-388-7
RTECS	WT1000000
RTECS class	Primary Irritant
Beilstein/Gmelin	3788739
Beilstein Reference	4-04-00-01524
EPA OPP	79046
Swiss Giftliste 1	G-4064
Canada DSL/NDSL	DSL
US TSCA	Listed
Austrailia AICS	Listed
New Zealand	Listed
Japan ENCS (MITI)	Listed
Korea ECL	Listed
Philippiens PICCS	Listed

Properties	
Formula	C18H41NO7S
Formula mass	415.66
Density	1.04 g/cm^3 (20 C)
Solubility in water	Soluble.

Hazards and Protection	
Storage	Storage temperature should not exceed 40 deg C. Product may become cloudy if stored below 0 deg C for prolonged periods. Avoid contact with copper and copper alloys. Keep in a cool place.

Hazards and Protection	
Handling	All chemicals should be considered hazardous. Avoid direct physical contact. Use appropriate, approved safety equipment. Untrained individuals should not handle this chemical or its container. Handling should occur in a chemical fume hood.
Protection	Eye - Wear chemical monogoggles. Skin - Wear gauntlet type nitrile rubber gloves, standard issue work clothes and chemical resistant safety shoes or boots. If splashes are likely to occur, wear disposable one-piece coated overall with integral hood.
Respirators	Not normally required. If risk of inhalation of mists/spray, wear half mask respirator with organic vapour cartridge and built-in particulate filter.
Small spills/leaks	In case of small spills, absorb or contain liquid with sand, earth or spill control material. For large spills, prevent from spreading by making a barrier with sand, earth or other containment material. Shovel up and place in a labeled, sealable container for subsequent safe disposal. Flush contaminated area with plenty of water.
Stability	No data.
Incompatibilities	Strong oxidizing agents.
Decomposition	Osides of carbon, nitrogen and sulfur.

Fire	
Fire fighting	Wear a self-contained breathing apparatus in pressure-demand, MSHA/NIOSH (approved or equivalent), and full protective gear. During a fire, irritating and highly toxic gases may be generated by thermal decomposition or combustion. Use agent most appropriate to extinguish fire.

(*Continued*)

Fire	
Hazards	Toxic vapors of triethanolamine and oxides of nitrogen may form in fire.
Combustion products	Toxic vapors of triethanolamine and oxides of nitrogen may form in fire.

Health	
Poison_Class	4
Exposure effects	
Ingestion	Exhibits a low to moderate order of acute oral toxicity. May irritate the upper digestive tract.
Inhalation	Inhalation of vapours or mists may cause irritation.
Skin	Exhibits a low order of acute dermal toxicity. Severe irritant.
Eyes	Severe irritant.
First aid	
Ingestion	Consult a doctor if large amount was ingested.
Inhalation	Move patient to fresh air. Monitor for respiratory distress. If cough or difficulty breathing develops, evaluate for respiratory tract irritation, bronchitis, or pneumonitis. Administer oxygen and assist ventilation as required. Treat bronchospasm with beta2 agonist and corticosteroid aerosols.
Skin	Flush with water.
Eyes	Flush with water; consult a doctor if irritation persists.

Transportation	
USCG CHRIS Code	DST

Zirconium Oxychloride	
• Zirconyl chloride • Chlorozirconyl • Dichlorooxozirconium • Zirconium chloride oxide • Zirconium chloride oxide • Zirconium oxide chloride	
Formula	$ZrCl_2O$
Structure	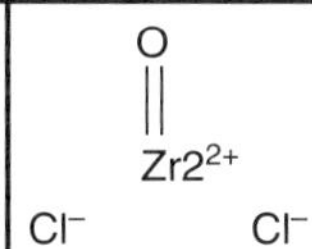
Description	White or yellow silky odorless crystals.
Uses	To make other zirconium compound, to precipitate acid dyes, to prepare high quality pigment toners, to improve properties of color lakes, has been used in prepn of body deodorants and anti-perspirant preparations.

Registry Numbers and Inventories	
CAS	7699-43-6
NIH PubChem CID	24373
EC (EINECS/ELINCS)	231-717-9
RTECS	ZH7700000
RTECS class	Tumorigen
Merck	12,10317
Beilstein/Gmelin	9649 (G)
Swiss Giftliste 1	G-3084
Canada DSL/NDSL	DSL
US TSCA	Listed
Austrailia AICS	Listed
New Zealand	Listed
Japan ENCS (MITI)	Listed
Korea ECL	Listed
Philippiens PICCS	Listed

Properties	
Formula	Cl2OZr
Formula mass	178.13
Melting point, °C	250 (decomposes)
Density	3.5 g/cm^3
Solubility in water	Soluble

Hazards and Protection	
Storage	Keep in a cool, dry, dark location in a tightly sealed container or cylinder. Keep away from incompatible materials, ignition sources and untrained individuals. Secure and label area. Protect containers/cylinders from physical damage.
Handling	All chemicals should be considered hazardous. Avoid direct physical contact. Use appropriate, approved safety equipment. Untrained individuals should not handle this chemical or its container. Handling should occur in a chemical fume hood.
Protection	Clothing contaminated with zirconium oxychloride, zirconyl acetate, or zirconium tetrachloride should be placed in closed containers for storage until it can be discarded or until provision is made for the removal of containment from the clothing.
Respirators	Use NIOSH/MSHA approved respirator appropriate for exposure of concern.
Small spills/leaks	Evacuate area and ventilate. Wear protective equipment. If required, use an inert absrobent. Sweep up and place in an appropriate container for disposal. Wash contaminated surfaces.
Stability	No data.
Incompatibilities	Reacts as acid to neutralize bases.

Fire	
Fire fighting	Use appropriate media to fight supporting fire. Contain runoff.

Health.	
Exposure limit(s)	IDHL: 500 mg/m^3
Poison_Class	4
Exposure effects	Experimental animals exposed to massive doses have developed CNS depression. Two individuals administered 50 milligrams of zirconium malate intravenously developed vertigo. Small fractions of zirconium were absorbed in female rats by the oral route, and the metal seemed to concentrate in the ovaries and produce hypervascularization. In mice, offspring of clams who received zirconium during pregnancy had long-lasting behavioral changes. Zirconium is found in newborn rats. In rats, zirconium appears in the breast milk.
Ingestion	Ingestion of certain zirconium salts may cause irritant or caustic effects.
Inhalation	Pulmonary abnormalities have not been demonstrated in workers with zirconium fume exposure for 1 to 5 years.
Skin	Dermal exposure to zirconium in topical poison ivy medications and deodorants has caused subcutaneous granulomas, probably due to a hypersensitivity reaction.
First aid	
Ingestion	Give large amount of water.
Inhalation	Move to fresh air.
Skin	Flush with water.
Eyes	Irrigate exposed eyes with copious amounts of tepid water for at least 15 minutes. If irritation, pain, swelling, lacrimation, or photophobia persist, the patient should be seen in a health care facility.

Transportation	
USCG CHRIS Code	ZCO_

Water

MSDS Number: W0600

1. Product Identification

Synonyms: Dihydrogen Oxide
CAS No.: 7732-18-5
Molecular Weight: 18
Chemical Formula: H_2O
Product Codes:
J.T. Baker: 4218, 4219, 6906
Mallinckrodt: 6795, H453, V564

2. Composition/Information on Ingredients

Ingredient	CAS No	Percent	Hazardous
Water	7732-18-5	100%	

No

3. Hazards Identification

Emergency Overview

Not applicable.

J.T. Baker SAF-T-DATA(tm) Ratings (Provided here for your convenience)

Health Rating: 0 - None
Flammability Rating: 0 - None
Reactivity Rating: 1 - Slight
Contact Rating: 0 - None
Lab Protective Equip: GOGGLES; LAB COAT
Storage Color Code: Orange (General Storage)

Potential Health Effects

Water is non-hazardous.

Inhalation:
Not applicable.

Ingestion:
Not applicable.

Skin Contact:
Not applicable.

Eye Contact:
Not applicable.

Chronic Exposure:
Not applicable.

Aggravation of Pre-existing Conditions:
Not applicable.

4. First Aid Measures

Inhalation:
Not applicable.

Ingestion:
Not applicable.

Skin Contact:
Not applicable.

Eye Contact:
Not applicable.

5. **Fire Fighting Measures**

Fire:
Not applicable.

Explosion:
Not applicable.

Fire Extinguishing Media:
Use extinguishing media appropriate for surrounding fire.

Special Information:
In the event of a fire, wear full protective clothing and NIOSH-approved self-contained breathing apparatus with full facepiece operated in the pressure demand or other positive pressure mode.

6. **Accidental Release Measures** Non-hazardous material. Clean up of spills requires no special equipment or procedures.

7. **Handling and Storage**

Keep container tightly closed. Suitable for any general chemical storage area. Protect from freezing. Water is considered a non-regulated product, but may react vigorously with some specific materials. Avoid contact with all materials until investigation shows substance is compatible.

8. **Exposure Controls/Personal Protection**

Airborne Exposure Limits: Not applicable.
Ventilation System: Not applicable.
Personal Respirators (NIOSH Approved): Not applicable.
Skin Protection: None required.
Eye Protection: None required.

9. Physical and Chemical Properties

Appearance: Clear, colorless liquid.
Odor: Odorless.
Solubility: Complete (100%)
Specific Gravity: 1.00
pH: 7.0
% Volatiles by volume @ 21C (70F): 100
Boiling Point: 100C (212F)
Melting Point: 0C (32F)
Vapor Density (Air=1): Not applicable.
Vapor Pressure (mm Hg): 17.5 @ 20C (68F)
Evaporation Rate (BuAc=1): No information found.

10. Stability and Reactivity

Stability:
Stable under ordinary conditions of use and storage.

Hazardous Decomposition Products:
Not applicable.

Hazardous Polymerization:
Will not occur.

Incompatibilities:
Strong reducing agents, acid chlorides, phosphorus trichloride, phosphorus pentachloride, phosphorus oxychloride.

Conditions to Avoid:
No information found.

11. Toxicological Information

——\Cancer Lists\——————————————

—— NTP Carcinogen ——
Ingredient Known Anticipated IARC Category

————————————————————

Water (7732-18-5) No No None

12. Ecological Information

Environmental Fate: Not applicable.
Environmental Toxicity: Not applicable.

13. Disposal Considerations Whatever cannot be saved for recovery or recycling should be flushed to sewer. If material becomes contaminated during use, dispose of accordingly. Dispose of container and unused contents in accordance with federal, state and local requirements.

14. Transport Information Not regulated.

15. Regulatory Information

——\Chemical Inventory Status - Part 1\——

Ingredient	TSCA	EC	Japan	Australia
Water (7732-18-5)	Yes	Yes	Yes	Yes

——\Chemical Inventory Status - Part 2\——

Ingredient	Korea	—— Canada— DSL	NDSL	Phil.
Water(7732-18-5)	Yes	Yes	No	Yes

——\Federal, State & International Regulations - Part 1\——

Ingredient	SARA 302 RQ	SARA 302 TPQ	SARA 313 List	SARA 313 Chemical Catg.
Water(7732-18-5)	No	No	No	No

——\Federal, State & International Regulations - Part 2\——

Ingredient	CERCLA	-RCRA- 261.33	-TSCA- 8(d)
Water (7732-18-5)	No	No	No

Chemical Weapons Convention: No TSCA 12(b): No CDTA: No
SARA 311/312: Acute: No Chronic: No Fire: No Pressure: No
Reactivity: No (Pure / Liquid)

Australian Hazchem Code: No information found.

Poison Schedule: No information found.

WHMIS:

This MSDS has been prepared according to the hazard criteria of the Controlled Products Regulations (CPR) and the MSDS contains all of the information required by the CPR

16. Other Information

NFPA Ratings: Health: **0** Flammability: **0** Reactivity: **0**

Label Hazard Warning:
Not applicable.

Label Precautions:
Keep in tightly closed container.

Label First Aid:
Not applicable.

Product Use:
Laboratory Reagent.

Appendix B

State Agency Web Addresses

Alabama, Geological Survey of Alabama, State Oil and Gas Board http://www.ogb.state.al.us/ogb/ogb.html

Arkansas, Arkansas Oil and Gas Commission http://www.aogc.state.ar.us/

Colorado, Colorado Department of Natural Resources, Oil and Gas Conservation Commission http://cogcc.state.co.us/

Illinois, Illinois Department of Natural Resources, Division of Oil and Gas http://dnr.state.il.us/mines/dog/index.htm

Indiana, Indiana Department of Natural Resources, Division of Oil and Gas http://www.in.gov/dnr/dnroil/

Kentucky, Kentucky Department for Energy Development and Independence, Division of Oil and Gas Conservation http://www.dogc.ky.gov/

Louisiana, Louisiana Department of Natural Resources, Office of Conservation http://dnr.louisiana.gov/cons/conserv.ssi

Michigan, Michigan Department of Environmental Quality, Office of Geological Survey http://www.michigan.gov/deq/0,1607,7-135- 3306_28607—,00.html

Mississippi, Mississippi State Oil and Gas Board http://www.ogb.state.ms.us/

Montana, Montana Department of Natural Resources and Conservation, Board of Oil and Gas http://bogc.dnrc.mt.gov/default.asp

New Mexico, New Mexico Energy, Minerals and Natural Resources Department, Oil Conservation Division http://www.emnrd.state.nm.us/OCD/

New York, New York Department of Environmental Conservation, Division of Mineral Resources http://www.dec.ny.gov/energy/205.html

North Dakota, North Dakota Industrial Commission, Department of Mineral Resources Oil and Gas Division https://www.dmr.nd.gov/oilgas/

Ohio, Ohio Department of Natural Resources, Division of Mineral Resources Management http://www.ohiodnr.com/mineral/default/tabid /10352/Default.aspx

Oklahoma, Oklahoma Corporation Commission, Oil and Gas Conservation Division http://www.occ.state.ok.us/Divisions/OG/newweb/og.htm

Pennsylvania, Pennsylvania Department of Environmental Protection, Bureau of Oil and Gas Management http://www.dep.state.pa.us/dep/DEPUTATE/MI NRES/OILGAS/oilgas.htm

Tennessee, Tennessee Department of Environment and Conservation, State Oil and Gas Board http://www.tennessee.gov/environment/boards /oilandgas.shtml

Texas, The Railroad Commission of Texas http://www.rrc.state.tx.us/index.html

West Virginia, West Virginia Department of Environmental Protection, Office of Oil and Gas http://www.wvdep.org/item.cfm?ssid=23

References

1. Interstate Oil and Gas Compact Commission "Hydraulic Fracturing, Texas Regulations Protect Surface and Ground Water"
2. Case Study for Well Integrity over a Full Life Cycle, Lloyd H. Hetrick, PE, CSP, EPA Technical Workshop #2 on Well Construction and Integrity, Theme #3 Mechanical Integrity, March 11, 2011
3. 40 CFR 144.3 Safe Drinking Water Act, definition for USDW
4. Texas Administrative Code Title 16, Part 1, Rule 3.7 and Title 16, Part 1, Rule 3.8 (b)

5. 43 CFR 3162.1 (a) Onshore Oil and Gas Order #2, Drilling Operations
6. U.S. Energy Information Administration "The Geology of Natural Resources" February 14, 2011
7. Pennsylvania Department of Environmental Protection "Methane Gas and Your Water Well" November 2009
8. The National Academies "Management and Effects of Coalbed Methane Produced Water in the United States" 2010
9. *New York Times* "Nature's Cheapest Fuel, The Natural Gas of the Delaware Valley Likely to be Utilized" March 15, 1886
10. Western New York Trails.com "Eternal Flame Falls – Chestnut Ridge Park, Orchard Park, New York" December 2010
11. American Geophysical Union Blogosphere "Natural Gas Seeps in Western NY" November 2010
12. Penn State University College of Agricultural Sciences "Gas Well Drilling and Your Private Water Supply" March 2010
13. Wikipedia, Oil Springs, Kentucky
14. Wikipedia, Oil Springs, Ontario Canada
15. Wikipedia, Seneca Oil Spring, NY
16. Texas Landmarks and Vanished Communities, Salt Creek and Brine Springs, TX
17. Oklahoma Administrative Code 165:10-3-1
18. Texas Administrative Code Title 16, Part 1, Rule 3.13 (a)(2)(C)
19. Pennsylvania Code Title 25 Chapter 78.83
20. Oklahoma Administrative Code 165:10-3-4 (c)(1)
21. Pennsylvania Code Title 25 Chapter 78.84 (a)
22. API Spec 5CT "Specification for Casing and Tubing" Eighth Edition 2006
23. NACE MR0175 / ISO 15156 "Materials for use in H2S Containing Environments in Oil and Gas Production" Second Edition 2009
24. API RP 5C1 "Recommended Practice for Care and Use of Casing and Tubing" Eighteenth Edition 2010
25. API RP 5C5 "Recommended Practice on Procedures for Testing Casing and Tubing Connections" Third Edition 2003
26. API RP 5A5 "Field Inspection of New Casing, Tubing, and Plain-end Drill Pipe" Seventh Edition December 2009
27. API RP 5B1 "Gaging and Inspection of Casing, Tubing, and Line Pipe Threads" Fifth Edition 2010
28. API RP 5A3 "Recommended Practice on Thread Compounds for Casing, Tubing and Line Pipe" Third Edition 2009
29. Department of Energy, Sandia National Labs "Euler Buckling of Geothermal Well Casing" Rechard and Schuler, April 1983
30. Texas Administrative Code Title 16, Part 1, Rule 3.13 (b)(2) (A) thru (G)
31. Texas Administrative Code Title 16, Part 1, Rule 3.13 (a)(1)

32. API RP 65 "Isolating Potential Flow Zones During Well Construction" Part 2, First Edition May 2010
33. API Spec 10A "Specification for Cements and Materials for Well Cementing" December 2010
34. API RP 10B (1 thru 6) "Recommended Practice for Testing Well Cements"
35. API Spec 10D "Specification for Bow-Spring Centralizers" Sixth Edition 2002
36. SPE 106817 "Evaluation of the Potential for Gas and CO2 Leakage Along Wellbores" Watson and Bachu, March 2007
37. Alberta Energy and Utilities Workshop "Factors Affecting or Indicating Potential Wellbore Leakage" Watson and Bachu, March 2007
38. API Technical Report 10TR1 "Cement Sheath Evaluation" Second Edition 2008
39. SPE 105193 "Improving Formation Strength Tests and Their Interpretation" van Oort and Vargo, February 2007
40. Texas Administrative Code Title 16, Part 1, Rule 3.17 (a) & (b)
41. 43 CFR 3162.5-2 Onshore Oil and Gas Order #2, Drilling Operations
42. API RP 90 "Annular Pressure Management for Offshore Wells" First Edition, August 2006
43. Pennsylvania Code Title 25 Chapter 78.88(e)
44. Texas Administrative Code Title 16, Part 1, Rule 3.14
45. Oklahoma Administrative Code 165:10-11
46. Pennsylvania Code Title 25 Chapter 78.91 thru 78.98
47. API Bulletin E3 "Environmental Guidance Document: Well Abandonment and Inactive Well Practices for U.S. Exploration and Production Operations" First Edition, June 2000
48. SPE 28349 "Issues and Techniques of Plugging and Abandonment of Oil and Gas Wells" Calvert and Smith, September 1994
49. SPE 140482 "State and Federal Regulation of Hydraulic Fracturing: A Comparative Analysis" Arthur, Hochheiser and Coughlin; January 2011

Bibliography

Fjaer, E. (2008). "Mechanics of hydraulic fracturing." Petroleum related rock mechanics. Developments in petroleum science (2 ed.). Elsevier. *p. 369.* ISBN 978-0-444-50260-5. http://books.google.co.uk/books?id=l6CfasFxhzYC&pg=PA369&lpg=PA369.

Laubach, S.E.; Reed, R.M., Olson, J.E., Lander, R.H. & Bonnell, L.M. (2004). „Coevolution of crack-seal texture and fracture porosity in sedimentary rocks: cathodoluminescence observations of regional fractures." *Journal of Structural Geology* (Elsevier) 26 (5): 967–982. doi:10.1016/j.jsg.2003.08.019. http://www.sciencedirect.com/science/article/pii/S0191814103001858.

„Definition of frac job." *Oilfield Glossary*. Schlumberger. http://www.glossary.oilfield.slb.com/Display.cfm?Term=frac%20job.

Office of Air and Radiation, U.S. Environmental Protection Agency (November 2010). „General Technical Support Document for Injection and Geologic Sequestration of Carbon Dioxide: Subparts RR and UU Greenhouse Gas Reporting Program." Office of Air and Radiation U.S. Environmental Protection Agency. http://www.epa.gov/climatechange/emissions/downloads10/Subpart-RR-UU_TSD.pdf.

Jack E. Whitten, Steven R. Courtemanche, Andrea R. Jones, Richard E. Penrod, and David B. Fogl (Division of Industrial and Medical Nuclear Safety, Office of Nuclear Material Safety and Safeguards) (June 2000). „Consolidated Guidance About Materials Licenses: Program-Specific Guidance About Well Logging, Tracer, and Field Flood Study Licenses (NUREG-1556, Volume 14)." US Nuclear Regulatory Commission. http://www.nrc.gov/reading-rm/doc-collections/nuregs/staff/sr1556/v14/#_1_26.

Anthony Andrews et al. (30 October 2009) (PDF). Unconventional Gas Shales: Development, Technology, and Policy Issues. Congressional Research Service. http://www.fas.org/sgp/crs/misc/R40894.pdf.

Bill McKibben (8 March 2012). „Why Not Frack?". *The New York Review of Books* 59 (4). http://www.nybooks.com/articles/archives/2012/mar/08/why-not-frack/.

David Biello (30 March 2010). „What the Frack? Natural Gas from Subterranean Shale Promises U.S. Energy Independence--With Environmental Costs". *Scientific American*. http://www.scientificamerican.com/article.cfm?id=shale-gas-and-hydraulic-fracturing.

(PDF) Chemicals Used in Hydraulic Fracturing (Report). Committee on Energy and Commerce U.S. House of Representatives. April 18, 2011. http://democrats.energycommerce.house.gov/sites/default/files/documents/Hydraulic%20Fracturing%20Report%204.18.11.pdf

Index

Also of Interest

Check out these other related titles from Scrivener Publishing

By the Same Author
Dictionary of Industrial Terms, by Michael Holloway and Chikezie Nwaoha, ISBN 9781118344576. This is the most comprehensive and in-depth dictionary of industrial terms available, a must-have for any engineer, scientist, technician, or operator working in industry. *NOW AVAILABLE!*

Other Related Titles from Scrivener Publishing
Bioremediation of Petroleum and Petroleum Products, by James Speight and Karuna Arjoon, ISBN 9780470938492. With petroleum-related spills, explosions, and health issues in the headlines almost every day, the issue of remediation of petroleum and petroleum products is taking on increasing importance, for the survival of our environment, our planet, and our future. This book is the first of its kind to explore this difficult issue from an engineering and scientific point of view and offer solutions and reasonable courses of action. *NOW AVAILABLE!*

An Introduction to Petroleum Technology, Economics, and Politics, by James Speight, ISBN 9781118012994. The perfect primer for anyone wishing to learn about the petroleum industry, for the layperson or the engineer. *NOW AVAILABLE!*

Ethics in Engineering, by James Speight and Russell Foote, ISBN 9780470626023. Covers the most thought-provoking ethical questions in engineering. *NOW AVAILABLE!*

Troubleshooting Vacuum Systems: Steam Turbine Surface Conden sersand Refinery Vacuum Towers, by Norman P. Lieberman. ISBN 978-1-118-29034-7. Vacuum towers and condensing steam turbines require effective vacuum systems for efficient operation.

Norm Lieberman's text describes in easy to understand language, without reference to complex mathematics, how vacuum systems work, what can go wrong, and how to make field observations to pinpoint the particular malfunction causing poor vacuum. *NOW AVAILABLE!*

Fundamentals of the Petrophysics of Oil and Gas Reservoirs, by Buryakovsky, Chilingar, Rieke, and Shin. ISBN 9781118344477. The most comprehensive book ever written on the basics of petrophysics for oil and gas reservoirs. *NOW AVAILABLE!*

Petroleum Accumulation Zones on Continental Margins, by Grigorenko, Chilingar, Sobolev, Andiyeva, and Zhukova. Coming in September 2012, ISBN 9781118385074. Some of the best-known petroleum engineers in the world have come together to produce one of the first comprehensive publications on the detailed (zonal) forecast of offshore petroleum potential, a must-have for any petroleum engineer or engineering student. *NOW AVAILABLE!*

Mechanics of Fluid Flow, by Basniev, Dmitriev, and Chilingar. Coming in September 2012, ISBN 9781118385067. The mechanics of fluid flow is one of the most important fundamental engineering disciplines explaining both natural phenomena and human-induced processes. A group of some of the best-known petroleum engineers in the world give a thorough understanding of this important discipline, central to the operations of the oil and gas industry.

Zero-Waste Engineering, by Rafiqul Islam, ISBN 9780470626047. In this controversial new volume, the author explores the question of zero-waste engineering and how it can be done, efficiently and profitably. *NOW AVAILABLE!*

Sustainable Energy Pricing, by Gary Zatzman, ISBN 9780470901632. In this controversial new volume, the author explores a new science of energy pricing and how it can be done in a way that is sustainable for the world's economy and environment. *NOW AVAILABLE!*
Formulas and Calculations for Drilling Engineers, by Robello Samuel, ISBN 9780470625996. The most comprehensive coverage of solutions for daily drilling problems ever published. *NOW AVAILABLE!*

Emergency Response Management for Offshore Oil Spills, by Nicholas P. Cheremisinoff, PhD, and Anton Davletshin, ISBN 9780470927120. The first book to examine the Deepwater Horizon disaster and offer processes for safety and environmental protection. *NOW AVAILABLE!*

Advanced Petroleum Reservoir Simulation, by M.R. Islam, S.H. Mousavizadegan, Shabbir Mustafiz, and Jamal H. Abou-Kassem, ISBN 9780470625811. The state of the art in petroleum reservoir simulation. *NOW AVAILABLE!*

Energy Storage: A New Approach, by Ralph Zito, ISBN 9780470625910. Exploring the potential of reversible concentrations cells, the author of this groundbreaking volume reveals new technologies to solve the global crisis of energy storage. *NOW AVAILABLE!*